SKIBO

Skibo Castle
(Photograph by Mrs. Andrew Carnegie)

SKIBO

JOSEPH FRAZIER WALL

New York Oxford
OXFORD UNIVERSITY PRESS
1984

Library of Congress Cataloging in Publication Data
Wall, Joseph Frazier.
Skibo.
Bibliography: p. 1. Skibo (Highland Region, Scotland)
2. Carnegie, Andrew, 1835–1919—Homes and haunts—Scotland.
3. Carnegie family. I. Title.
DA890.S6W35 1984 941.1'65 83-26763
ISBN 0-19-503450-3

The author acknowledges the kind permission of the publisher for the use of selections from *Louise Whitfield Carnegie: The Life of Mrs. Andrew Carnegie* by Burton Hendrick and Daniel Henderson. © Copyright 1946, Hastings House, publisher.

Printed in the United States of America

Printing (last digit): 9 8 7 6 5 4 3 2 1

For Margaret Carnegie Miller
Gracious and Good Lady of Skibo
This is her castle—her book.

Acknowledgments

The contents of a book are the sole responsibility of the author, but in the act of creation, the author must share pride of parentage with many other people. I gratefully acknowledge those who have been partners with me in the writing of this history of Skibo:

Alan Pifer, President Emeritus of the Carnegie Corporation of New York, who first suggested to me a history of Skibo; Sara L. Engelhardt, Secretary of the Corporation; and the Corporation itself for having made research funds available to me.

The Thomson family: Gordon Thomson (Lord Migdale), and his five children, Betty Thomson Milligan; Margaret Thomson; Louise Thomson Suggett and her husband, Gavin; Mary Thomson and William Thomson. Their memories of Skibo enriched these pages, and their care and patience in reading the manuscript provided the necessary corrections of factual errors in the text.

The grandchildren of Margaret Carnegie Miller: Linda Thorell Hills, who also shared with me her experiences at Skibo and gave the manuscript the critical review it needed, and Roswell and Kenneth Miller. I am particularly indebted to Kenneth for his making available to me the diaries of his father.

Geoffrey Lord, Secretary of the Carnegie United Kingdom Trust, and Fred Mann, Secretary of the Carnegie Dunfermline Trust. They both welcomed me royally to the Royal Burgh of Dunfermline and provided me with many useful documents and photographs.

The many others who knew and loved Skibo and who helped me to understand its peculiar magic: Mrs. Nan Carnegie Rockefeller, Andrew Carnegie's niece, who visited Skibo in its days of glory; Mr. and Mrs. Arthur D. L. Robertson of the Dunfermline Trust and Baroness Elliot of Harwood, of the United Kingdom Trust; Mrs. Robert Grant of Dornoch and the Reverend James Simpson of Dornoch Cathedral; Reathel Odum, Mrs. Miller's companion during the last four summers of her residency, whose diary of those years gave to me an invaluable insight into the final phase of the Carnegie-Miller era; Ruth Adams, Mrs. Miller's good friend and faithful amanuensis; and the many employees of the Skibo estate, both past and present, with special thanks to Mr. and Mrs. Bobby Edgar and the late, great gamekeeper, Harry Blythe, whose pedagogical skills were so remarkable as to make understandable, even to this ignorant Iowan, the intricacies of the sports of field and stream in the Highlands of Scotland.

Sheldon Meyer and Leona Capeless of Oxford University Press, who have so expertly guided me through the process of getting published for many years.

Above all, I am indebted to Margaret Carnegie Miller. In every respect, this is her book, and my greatest personal reward has come from the knowledge that I had her encouragement and her friendship while serving as scribe for this history of Skibo.

Grinnell, Iowa J. F. W.
December 1983

A sad postscript to the above must be added to note the death of Lord Migdale on 30 December 1983. He loved the lands of Skibo and Ospisdale, and it is a matter of great personal regret to me not to be able to present to him this account of Skibo's history to which he made his own very special contribution.

Contents

SKIBO

I

Dark Beginnings for Sunny Skibo

?–1745

The traveler, driving along north Scotland's Route 9A from the village of Bonar Bridge at the head of the Firth of Dornoch toward the ancient cathedral town of Dornoch, some fifteen miles distant, can easily pass Skibo Castle without being aware of its existence. For unlike most of Scotland's famed castles, Skibo does not flaunt itself openly to the land traveler. It does not dominate the landscape as does its neighbor, Carbisdale Castle at Culrain, nor cause the passing traffic to hesitate with awe in its flow as do the romantic ruins of Urquhart on the promontory overlooking Loch Ness farther to the south. No notice marks the entrance to Skibo. Only a pleasant little gate house by a break in the stone wall, and farther on, some rather imposing farm buildings give a hint that perhaps something significant lies beyond the birch, beech, and pine trees that screen the interior from the road.

But for the traveler coming by boat up Dornoch Firth past Ferry Point in Cambuscurrie Bay, Skibo shows itself in all of its majesty, with its towers, crenellated parapets and terraces looking as if they had been but recently scooped up and moulded out of the white sands of the Dornoch beaches by children of some super race of titans. The uninformed boat riders might well gasp, "What in the world is that?", while excitedly pointing their cameras at the great mass of white sandstone, only to be told by their native guide, "Why, that's old Andra' Carnegie's summer place. That's Skibo."

Skibo. It is a most euphonious name, pleasing to the ear and far more accommodating to the non-Gaelic tongue than so many of the neighboring place names of Achinduich, Claes na Sinnaig and Loch a' Ghobhair. The very meaning of Skibo is lost in the distant, linguistic past. The final syllable, "bo," would indicate that Skibo is a word of Old Norse origin, for "bo" is the Scandinavian equivalent of the French "chez," meaning place or home of. With this as a clue, the amateur onomatologist could translate from Old Norse the two syllables and easily deduce that Skibo meant "the place of the sticks of wood." Such an analysis of the component parts has its own inner logic, for one can imagine that the first Norse invaders of this land of the Picts might well have built a wooden palisade around this pleasant and sheltered landing spot to protect their camp and moorage from stealthy, nocturnal forays of the savage Picts living in their crude huts up on the moors.

A very plausible explanation for the name Skibo, but unfortunately the origins of a word are seldom so simply determined as this literal translation from the Old Norse of the present-day syllables would indicate. Skibo is an ancient place, and its name has undoubtedly been much changed and corrupted over time. Professional linguists will not accept so pat an answer, nor on the other hand can they reach any consensus among themselves as to Skibo's original meaning. Some insist that "ski" is a corruption of the old Teutonic word "skif," and that Skibo means quite simply, "the place of ships."[1] This interpretation would also have its historical validity. But then what is one to make of the fact that in the charter of 1225, Skibo Castle sports a quite different spelling? It is Schytherbolle–not Skibo–Castle that is assigned to the

Bishop of Caithness by the Earl of Sutherland in that ancient charter. If this is indeed the original place name, then Skibo assumes an entirely different meaning. One must look beyond the Old Norse dictionary for its origin. "Bolle," as a form of the Norse "bo," still remains as a land or place designation, but the first syllable, "schy," is Irish Celtic. Pronounced "shee" in Celtic (which would also be the pronunciation for "ski" in Norse), "schy" or "schie" in Gaelic means "fairy hill," and also interestingly enough, "peace," since the Irish fairies apparently were the only people of peace in that sad, war-ravaged green island.[2] We now have the possibility that Skibo once indicated fairyland, or a place of peace. To many of its later owners, most notably Andrew Carnegie, these two possibilities would be the most appropriate meanings of all for their beloved Skibo.

But if Skibo originally meant "land of peace," it was only the land, certainly not its early inhabitants, that gave reality to the name. Nature, to be sure, has been kind to the shores of the Dornoch Firth. Protected by the Ross hills to the south, the high moors to the north, and that long ridge of sand lying across the mouth of the firth called Gizzen Briggs–which ancient legend claims was built by the fairies–this short far-northern firth of Scotland rarely receives the full blast of the cruel North Sea, nor do its lands get the full effect of the northwestern storms that come down from the Arctic to chill most of Scotland. Here along this sheltered bay, the air is soft and mild during many months of the year. Rhododendron has been known to bloom in January in this land that lies as far north as Churchill does on the west central shores of Hudson's Bay. In a country not famed for sunbathing, Dornoch parish is an easy victor in the contest for having the most days of sunshine in all of Scotland. Nature has done its utmost, with the geographic material it has to work with, to make this small spot a fairyland of peace.

But as so often is the case elsewhere, here in Skibo, man

and nature were seldom in tune. History has been inclined to turn Thomas à Kempis's famous maxim—"man proposes, but God disposes"—upside down. God may have proposed Skibo to be a sunny land of peace, but man, during most of recorded history, disposed otherwise. Its gently sloping white beaches have been as stained with blood as the wild terrain of Killiecrankie Pass. Skibo's soft spring air has been as punctured with the same harsh sounds of metal striking upon metal as the whistling winds that blow over the heights of Cawdor.

We know very little about the first inhabitants of the lands that would eventually make up the estate of Skibo. Perhaps in their huts and round stone towers called broches, these prehistoric people may have lived in peace with one another, but the contents of their secreted hoards and burial cairns which have been unearthed on the Skibo estate would indicate that these people of the Bronze Age, like the later inhabitants, also lived and died by the sword, or, as in their case, by their flat bronze axes. We do not know where these people came from. It may be that they were people of Mediterranean origin who came by their crude boats through the Pillars of Hercules up the long west coast of Portugal, France, and Britain past the western isles of the Hebrides for reasons unknown to find a haven in these remote lands of ultima Thule. Nor do we know what became of them. Were they the remote ancestors of the later Picts, those naked, tatooed savages who were to occupy all of north and central Scotland and become objects of fascination and fear for the later invading Romans? Probably not. Most ethnologists believe that the later dominant Picts were of a quite different race and culture from the Bronze Age people. It is believed that the Picts, possibly coming out of the Austrian valleys, were a part of the Celtic linguistic family, but different in physical appearance and in social customs from their kinsmen of Wales, Brittany, and Ireland. If they were indeed Celts, these Picts must simply have absorbed the few remaining Bronze Age people they had not slain, and then established for themselves a rough hegemony over the highlands of Scotland, from the Shetlands to the Firth of Forth.

Having ended all indigenous opposition, perhaps the Picts of Dornoch Firth were then able to live in peace in this peaceful land, but if so, not for long. They soon faced intrusion into their lands by other Celts—the Britons, who came into Strathclyde, and the Irish Celts, called Scots, who left Ireland to form a beachhead in the western isles and on the mainland in Argyll. It was only invasion by Roman legions who pushed north into Scotland in the first century A.D. that delayed an intra-Celtic struggle for the control of Pictland.

The Romans through three centuries of struggle never managed either to subjugate or to pacify these Celtic Caledonians of the north. Only in the border country, where under Emperor Hadrian the Romans built their great wall, was the Pax Romana brutally effective. But it was a bitter and arid peace. In the words of the Roman historian Tacitus, speaking through his Pictian spokesman Calgacus, "They make a desert and they call it peace."

The Roman soldiers under their commander Agricola got as far north as river Tay, and a century later under the direct command of the Emperor Severus they may have reached the Firth of Moray. Dornoch Firth, some thirty miles further north, never saw a Roman ensign, but the Picts living here knew well the costs of the Roman wars. Their men who went south to fight the invaders died in such numbers that it was reported that the Romans retreated back to the safety of their walls across streams bridged with the dead bodies of Picts.

As the Roman legions retreated back out of Caledonia and then finally out of Britain entirely, they were succeeded by a new kind of invasion into the land of the Picts, that of the Christian missionaries. Although historians differ, it may possibly have been St. Ninian, expanding his mission northward from his home base in Galloway, who first made contact with and converted some of the Picts in the fifth century A.D. There is no dispute over the fact that Columba, a century later, met the king of the Picts, Bridei, in the Highland wilderness near Inverness and won this great heathen king over to Columba's true faith. Columba, who took his name from the Latin word for dove of peace, met the heavily armed Bridei wearing no armor, only a monk's simple tunic, and bearing no sword, only the crucifix. Columba was a disciple of the Prince of Peace but, unfortunately, Christianity brought no peace to the strife-ravaged Highland of the Scots and the Picts, or to the shores of Dornoch farther to the north and east.

It was Finnbar, the fair-haired one, who first brought the Christian message from St. Ninian's College in far-distant Solway to the lands that surround Skibo. Here in the little Pict village where the Dornoch Firth meets the North Sea, he built the first Christian church, established a monastery for his coarse-woolen-robed monks, and preached the doctrine of brotherly love. Love and peace—but there was no peace in Dornoch, or anywhere else in Christendom. Men would die over the proper date for celebrating Easter—the Resurrection of Life. Old women would be hanged on Gallows Hill in Dornoch as heretic witches of the true faith. Finnbar's church would be torn down and, because they professed Ninian's version of Christianity his monks would be driven from town by those who called themselves disciples of Columba, that gentle dove of peace. Columba's Culdee monks would in turn be driven out of Dornoch by a reassertion of Roman power, emanating this time not from the emperor but from the bishop of Rome. Christianity had provided still another excuse for man's inhumanity to man. If questions of state did not provide enough reasons for war, questions of faith most certainly could.

For two centuries after the conversion of the Picts to Christianity, men did not need doctrinal disputes to justify their taking up the shield and the sword. The invasion of the Norsemen made even the dispute over a long or short Lenten period pale into insignificance. From the moment the Vikings first beached their long graceful galleys on to the sheltered Dornoch sands in the late ninth or early tenth century, the lands surrounding Skibo were to witness little else but war. It was these Norse invaders who gave to this entire far northeastern section of Britain its present-day name of Sutherland, for it was indeed to them, coming down from Norway via the Shetland and Orkney islands, a Sudrland–their southern land of occupation.

Here around Dornoch the great Jarls of Norway, Thorfinn and Thorkill, Sweyn and Sigurd, built their castles at Borve and Kylesku, and most probably some kind of stronghold at Skibo–not yet imposing enough to merit the title of castle. Here they waged war on the Picts, forcing them into alliance with their ancient foes and kinsmen, the Scots of Dalriada and the Britons of Strathclyde. The Norse slew the Celts and were in turn, like Olaf, the great Danish chief, slain and buried in the meadows and uplands that border Dornoch Firth.

Here too, the Norse interbred with their defeated and their conquering foes, producing that special breed of Briton known as the Highland Scot, mostly Celtic in ancestry but with enough Norse blood to distinguish him from his darker and smaller Gaelic kinsman of southwestern Scotland.

The mixing of these people did not, however, bring peace to the land. In the long centuries of struggle to create a modern nation-state, beginning with the acquisition of all of Pictland by Kenneth mac Alpin, a Scot of Dalriada (Argyll), in the mid-ninth century and culminating with Robert the Bruce's victory over the English at Bannockburn in 1314, Celts continued to fight Celts, and together to fight the English, the Northumbrians, and the increasingly infrequent Norse invaders.

It was in the middle of this long period of building an independent Kingdom of Scotland that a new strong family emerged in Sutherland who became protective possessor and donor of the lands of Skibo. Hugo Freskin of the de Moravia family may have been one of the progeny of the Flemish artisans whom the Normans imported into Britain in the century following their successful invasion of England in 1066, or he may have been descended from one of the Norse invaders who settled into the lands around the Firth of Moray, as some historians insist. In any event, as head of the de Moravia family, he was granted vast estates in Sutherland by the Scottish king, William the Lion, soon after that monarch had brought all of the lands of the far northeast under the central control of the Scottish crown. Freskin, in turn, subinfeudated large portions of his Sutherland bounty to his many Moravian relatives.

Among the recipients of his largesse was his cousin, Gilbert de Moravia, the archdeacon of Moray. To Gilbert, Freskin in 1186 bequeathed "The land of Skelbo in Sutherland and of Fernebuchlyn and Invershin, with the land of Sutherland towards the west which lies between these lands before-named and the boundaries of Ross."[3] It was a handsome bequest, containing many thousands of acres, including those belonging to the estate of Skibo, although as yet, Skibo was not of great enough significance to be specifically designated in the grant.

In selecting his cousin Gilbert to receive one of the larger grants, Hugh Freskin had chosen a kinsman whose star was clearly in the ascendancy. In 1222, Gilbert was named bishop of Caithness, a diocese that included the lands and people in both the shires of Caithness and Sutherland. It was a position of power, for like the bishop of Durham, the bishop of Caithness was recognized by both the crown and the church as a Lord Spiritual and a Lord Temporal. Both cassock and armor were appropriate garb for this bishop, and Gilbert was just the man to wear both with great authority.

One of Gilbert's first acts was to build upon the ruins of the old Culdee church in Dornoch a new cathedral befitting the see of a powerful bishopric. Because of the vast amount of territory within his jurisdiction, Gilbert maintained two residential palaces. The first was in Scrabster at the extreme end of Caithness shire on the Pentland Firth, where Britain turns its northern face to the open sea and to the Orkney and Shetland Isles beyond. For his second and more frequented place of residence, Gilbert chose the sunny land of Schytherbolle at the southernmost limit of his bishopric, only four miles from his cathedral in Dornoch. For this residency, he may have made use in part of whatever remained of the existing stronghold built by the Norse two centuries earlier, but upon these ruins, he built an imposing fortress-like castle, with a high round keep, not unlike the ancient broches which the precursors of the Picts had built in this land many centuries earlier. Below the ramparts, on the several terraces descending to the firth, Gilbert's attendants planted their vegetable gardens, their orchards and flower beds. Skibo—or Schytherbolle, as it was then called—at long last had achieved a position of prominence. In the renewal of the grant of subinfeudation to the bishop of Caithness by Hugh Freskin's son, William, the newly created Earl of Sutherland, Schytherbolle in 1225 was specifically mentioned as a castle, which by then had become Gilbert's favorite possession and his chief place of residence.

It had undoubtedly been the hope of both the Scottish monarch and the Roman Catholic Church that in creating the bishopric of Caithness which would include both the shire of Sutherland and that of Caithness, and by giving to its bishop powers both temporal and spiritual, an authority would be established in these far northern reaches of the church and state that would keep both God's and the king's peace. Under Gilbert, the plan worked. His was such a commanding presence throughout both shires that during Gilbert's lifetime the rival clans and the two proud earls of Sutherland and Caithness kept within their territorial limits in a watchful, if ever-suspicious détente. Schytherbolle had at last found the peace that its name proclaimed, and this peace extended to the northernmost point of Britain. "Holy Gilbert" was the undisputed master of the region, a man of God whose power was respected by the nobles and whose godliness was adored by the peasants and burghers. Miracles even began to be attributed to his holy presence.

But Bishop Gilbert proved to be only mortal after all. While in temporary residence at his palace in Scrabster in the summer of 1245, he died. His body was brought the long way back by sea to the cathedral in Dornoch to be buried under the floor of the nave. Along with his body was buried the peace that Gilbert had maintained.

The bureaucracy in Rome moved quickly to canonize Gilbert, the last native Scotsman who was to receive this highest honor of the Church. In the calendar of saints, April 1st was assigned as Saint Gilbert's feast day. Perhaps Rome had hoped that with sainthood Gilbert's power would extend beyond the grave and his peace would be kept in northern Scotland. If so, it proved to be a vain hope—a cruel April Fool's Day joke.

For the next half-millennium, Sutherland and Caithness were to be ravaged by war as the earls of those two shires continued to struggle for predominance in this northern land, and their even more war-like allied clans, the Mackays and Macleods, did battle with the Sinclairs and the Mackenzies. Conflict in the Highlands and on the coastal plains became fragmented and diffuse but no less intense than it had been when the inhabitants had been united in their opposition to the Roman and Norse invaders. Once again war became an accepted way of life and for many Highlanders a certain cause of death.

Frequently during these five bloody centuries of conflict, the local feuds would be merged into the larger battlefield of dynastic struggle for succession to the Scottish and English thrones–the royal Stewarts pitted against the last of the English Tudors, and later the Jacobite Stuarts vainly seeking retribution from their Hanoverian kinsmen. Here in this small kingdom of Scotland there would be glory and gore aplenty–enough to suffice the ballad singers and storytellers for ages to come: stately and beautiful Mary, Queen of the Scots, repeatedly offering in song and story her lovely head to cruel Elizabeth's axeman in Fotheringay Castle, or Bonnie Prince Charlie, forever sailing into exile beyond the mists of Skye. Here was the stuff of high drama, here the inspiration for all future lost causes to find victory in legend after having known defeat in history. Scotland in these centuries knew much romance, but little peace. The ancient Celtic fairies must have sealed themselves off deep within some subterranean cave for there was no harmony of man with nature on their Skibo's sunlit shores.

Nor did Christianity provide any solace of peace in these years. It proved to be no solution but rather a major source of Scotland's problems. The priests who were to succeed Gilbert as bishops of Caithness were often men of power and authority but, unlike him, they were not, successful saints of peace. Too often they would side with either the Earl of Sutherland or the Earl of Caithness rather than, as Gilbert had done, keep the earls and their minions apart. Each successive bishop continued to reside in Gilbert's fortress-like castle at Skibo even after the Church had built a residential castle for the bishop directly across the road from the cathedral in Dornoch. Skibo had attractions which Dornoch Castle could not offer, not the least of which being privacy and separation from the daily demands of diocesan office that the four miles' distance provided.

By the mid-sixteenth century, however, the Roman Catholic Church was in serious trouble in Scotland as it was elsewhere throughout western Europe, and the good bishops at Dornoch could no longer enjoy the luxury of isolating themselves from the immediate demands of the cathedral. The trouble had first begun on Halloween night in 1517 in far-distant Germany when a young Dominican monk, troubled by doubts of his own personal salvation and angry at the continuing monetary demands of the bureaucracy in Rome, nailed to the church door in Wittenberg a proclamation of ninety-five theses attacking the current practices and, more seriously, some of the fundamental doctrines of the Mother Church of Western Christendom. With each blow of his hammer on the church door, Brother Martin smashed the unity of eggshell fragility that for a thousand years had bound together all of medieval Europe north of the Danube and west of the Vistula. Humpty Dumpty once smashed could never be put together again. With his ninety-five theses, Luther had let loose all of the pent-up doctrinal differences that the universal Catholic Church of Rome had long kept suppressed. He had put St. Augustine over St. Peter, the Bible over the priest and his sacraments, the national prince over the international pope. Luther overnight turned potential reform into real revolution. Minor isolated protests had now become concentrated, capitalized and institutionalized into the Protestant Reformation. Religion would no longer deter but instead aid and abet the nationalistic and dynastic struggles that had long threatened European unity.

In 1536, Henry VIII, whom the Pope but a decade before had proclaimed Defender of the Faith, forced a subservient Parliament to recognize him as head of the Anglican Catholic Church and quickly the flames of Protestant revolt were spread across the border into the Lowlands of Scotland. In 1545, exactly three hundred years after Saint Gilbert had been laid in his grave in Dornoch Cathedral, a dour, grim-visaged ex-priest, John Knox, began to preach his sermons of predestined doom and unmitigated gloom in the ancient university town of St. Andrews. With those sermons, the old romantic Scotland of Celtic fairies, of graceful abbeys and noble cathedrals was put away forever.

In that same year of 1545, the then-ruling bishop of Caithness, Robert Stewart, ironically observed the tercentenary of the death of Saint Gilbert by ceding away Gilbert's most cherished possession. Bishop Robert gave in perpetuity Gilbert's residential castle of Skibo to John Gray, who had already received much of the land surrounding Skibo from the Earl of Sutherland. In so doing, Bishop Robert was making a tactically wise decision. In these days of trouble it would be well for the bishop to live in close proximity to his cathedral. There would perhaps be less likelihood of the cathedral-burning, window-breaking, and icon-smashing that had already destroyed the abbeys of Melrose, Dryburgh, and Jedburgh and the great cathedral of St. Andrews in southern Scotland if Bishop Robert and his retinue were in residence directly across the street from Dornoch Cathedral.

Unquestionably, Robert was also motivated to give up the ancient home of the bishops of Caithness in order to strengthen an alliance with a powerful family in the region as the Protestant revolt continued its inexorable crawl northward. And the Grays were a powerful family, in close association with the Earl of Sutherland to whom they owed fealty for their extensive land holdings in the southern parishes of Sutherland. Moreover, John Gray had a very personal reason to be grateful to and supportive of the Church. His father, Sir William Gray, had been chaplain of the vicarage of Rosemarkie, and as kirk celibate, had never married. He had, nevertheless, sired two sons, John and William. In 1539, the Church had obligingly provided both John and his brother, William, with letters of legitimation which gave them legal standing before both church and state. Owing his very legitimacy of birth as well as his castle, his revenue from former Church lands, and his title, Constable of Skibo, to the Roman

Catholic Church, John Gray was not likely to be found among the Protestant rebels.[4]

So in 1545, the bishopric of Caithness had obtained for itself a powerful defender and Skibo had been given a new master. John Gray proudly designed a new Constable's banner which bore "a lion rampant within a bordure engrailed charged with eight thistles, within a shield; crest—an arm erect grasping a heart; motto—"Constant."[5] The Grays' constancy, however, pertained only to their castle and their title. During the next two hundred years they would alternately back the Stewarts and the Hanoverians as the fortunes of war changed. They did not even remain true to the Church that had given them their castle and their title. John Gray's grandson and namesake would become the first Protestant Dean of Sutherland and Caithness and would serve as Presbyterian minister for the parish of Dornoch for thirty years. "Survival" would have been a better motto for the Grays than "Constant," for although they fought and died for all of the many causes that Scotland provided in these years, they also married, bred—and survived as the proud Constables of Skibo.

The Grays, to be sure, were merely reflecting the many changes that Scotland itself suffered during these two centuries. Although the old faith died slowly in the remote glens and straths of the Highlands, ultimately even there Geneva and John Knox won out over Rome and Mary Stewart. Scotland, once the last northern outpost of Catholicism would become the most Calvinistic of all European states. Feast days of celebration yielded to fast days of necessity. Christmas became simply December 25th, and Sunday became the Scottish Sabbath, no longer a happy day of recreation but a somber day of reformation and puritanical introspection. The Auld Alliance which had long tied Scotland with France was forever dissolved when James Stewart, Mary's son, went south to London to claim the English throne—Mary Stewart's final victory by default over Elizabeth Tudor—but a century later, the last of the Stewart monarchs, Queen Anne, would preside over the final extinction of the Scottish nation. Scotland in 1707, by the Act of Union, became simply another region of Great Britain, like Wales or Cornwall or Northumberland, and the ancient Stone of Scone, that rock which legend said had once served as Jacob's pillow when he dreamed his dream of angels ascending to heaven and which had for centuries served as the coronation seat for Scottish kings, was now firmly ensconced with the English coronation throne in Westminster Abbey.

But a few things did remain constant during these centuries of change: war among the noblemen and clergy, and poverty among the villagers and crofters. Always a poor country, Scotland became even poorer in these years as the state demanded more tax monies for war, and the wool trade demanded more land for the sheep. Dispossessed of their land, the crofters drifted into the cities of the Lowlands to find miserable shelter and slow starvation in the narrow closes of Edinburgh and the rapidly growing slums of Glasgow.

> Ill fares the land, to hastening ills a prey,
> Where wealth accumulates, and men decay.
> Princes and lords may flourish, or may fade,
> A breath can make them, as breath has made:
> But a bold peasantry, their country's pride,
> When once destroy'd, can never be supplied.[6]

So did the Irish poet, Oliver Goldsmith, write about the "Sweet Auburn" of his youth, now a deserted village, but his lament would be as meaningful to the villages and crofts of Sutherland, Scotland, as it was to County Westmeath, Ireland.

Ultimately, the Gray family could not remain constant even to Skibo and its banner. In 1745, five hundred years after Gilbert de Moravia, the saintly bishop of Dornoch, had said his last farewell to his beloved Skibo, and two hundred years after John Gray had taken possession of the castle and the title of Constable of Skibo, his several times great-grandson, young Lieutenant Robert Gray, burdened by debt and unable to meet the financial obligations owed to his two demanding half-sisters, surrendered his inheritance to pay off his creditors in Edinburgh. Leaving Skibo forever, he went off to England to serve in His Majesty's, King George's, army.

In that same commemorative year of 1745, Bonnie Prince Charlie landed on Eriskay in the Outer Hebrides with seven companions, and being joined by a few hundred supporters of the Cameron and MacDonald clans, raised his standard at Glenfinnan. It was Scotland's last try for the Romance, the last hurrah for glory, and it was soon over. On Culloden Moor on 16 April 1746, the dream finally ended. And so they were all gone, the Stewarts gone from the Highlands, the bishops long gone from their old cathedral in Dornoch, and the Grays from their fortalice in Skibo.

There would be other wars in the centuries to come—the many wars which would take the young men from Dornoch and the little villages of the Skibo estate: Clashmore and Swordale, Spinningdale and Bonar Bridge. Lieutenant Robert Gray, now a major in the 55th Foot Regiment, would himself die in April 1776, three thousand miles from Skibo in a place called Staten Island in New York harbor. But never again would the beaches of Dornoch Firth or the high moors above Skibo resound with the clash of arms. The Celtic fairies, daring the curse of Calvin, could emerge once again from their subterranean retreats to dance on the hills, frolic in Fairy Glen, and build ever higher their Gizzen Briggs, those bridges of sand that lie like a protective shield across the mouth of the Firth. Skibo at last could be true to its old Gaelic name—Schytherbolle—a fairy land of peace.

II

Skibo Tamed and Domesticated

1745 – 1895

When Lieutenant Robert Gray in 1745 surrendered his lands and titles and went off to England to serve in the British army, the estate of Skibo which he left to his creditors consisted of the following:

> The townes and lands of Skibo, Castell Maynes of Skibo, Castell of Skibo, Mickell Swordaell, Tulloch, Migdaill, Litill Creich, Litill Swordaell, Cuthill, Cuthilhaldaill, Alistie, Steanford, Riengollach, Rienrich, Rienmamie, and Culnara, the heritable office of Constabularie of the said Castell of Skibo, Milns of Skibo and Migdale, Salmond fishings of the Bonar, Litill Creich, and Litill Swordaill, Townes and lands of Alistie, Auchvaich, and Ferrietown, passage of ferrieboat on the shoar of Portnaculter, Salmond fishings in the water of Poltchurich, Ferrieness and Ardnakalk northmost of the twa milns of Siddera.[1]

It was an impressive list of real property and designated privileges: towns, farms, mills, castle, ferry boats, and the right to salmon fishing in specified areas—a right which in Scotland was, and is, an assignable property right quite distinct from the ownership of the stream which the salmon may inhabit. Unfortunately, the centuries of warfare had exacted their toll upon the land as well as upon the people. Many of the crofts had lost their tenants, and the land had been reclaimed by the gorse, the rough brush, and the heather. The small villages had fewer inhabitants than in the fifteenth century, and the mills, if they ran at all, no longer ran at full capacity. The castle itself was in a bad state of maintenance, with its thatched roof partially gone and with only a few rooms still habitable. This ancient residential seat of the bishops of Caithness was in nearly as deplorable a state as the bishops' cathedral itself. Dornoch Cathedral, after being thrice burned, twice by the Calvinists and once by the Mackays, who supported Charles I, now stood as a ruined and sorry testament to Scotland's tragic history. Its roof was gone, but the walls of its great central steeple still stood, pointing like an accusing blunt finger to the heaven above. Within this general scene of desolation, it is hardly surprising that Gray's property, impressive as it may have been in its listing, could not fully satisfy the demands of his creditors.

For the next several decades, the ownership of the Skibo estate passed from creditor to creditor like a bad debt. The first owner after Robert Gray was, not surprisingly, Gray's Edinburgh lawyer, Sir Patrick Dowall. Dowall obtained Skibo and its lands from Gray for a price considerably below that of their appraised value, and then prorated to Gray's creditors their share of that agreed upon price. Dowall may have had plans to use Skibo as a country home, but he resided in Edinburgh and had neither the time, the money, nor the motivation to give the necessary attention to a run-down estate in far northern Scotland. After only six years of ownership, Dowall, at a good profit one assumes, transferred Skibo to his nephew, George Mackay. The latter was a man of prominence in Sutherland County, a member of Parliament, and a younger son of Lord Reay, the current chief of the Mackay clan. The Mackays had long battled with the earls of Sutherland for control of Dornoch and its surrounding territory.

George Mackay, unlike Dowall, took a keen interest in this ancient estate upon whose lands his ancestors had so frequently fought and died. When he was not in London attending sessions of Parliament, Mackay devoted most of his time attempting to restore the farms of Skibo to their former productivity. He even made a start in reforesting the denuded hills that lay behind the farmlands. But Mackay made no real attempt to make the old castle habitable. It remained in its sad, dilapidated state. George Mackay instead built two Georgian houses on the south and west of the castle keep to house his family and servants.

Although he came from a wealthy family, George Mackay soon discovered that with all of these improvements, Skibo was demanding more than he could give. Like his predecessor Robert Gray, Mackay found himself heavily in debt. It was consequently with considerable relief that he found someone with wealth enough to pay the price he wanted and needed to extricate himself from the financial morass in which Skibo had entrapped him. In 1756, just five years after he had taken over the estate with high ambitions for its improvement, George Mackay was obliged to sell out to William Gray.

Gray was a native of Sutherland County, who after having married Janet Sutherland, had gone off to Jamaica to seek and to find his fortune as a sugar planter. Upon returning to Scotland in 1756, Gray was delighted to be able to purchase from Mackay the ancient ancestral home of the John Gray family with whom he claimed a remote relationship. So within a little over ten years after Robert Gray had surrendered his inheritance, there now appeared upon the deed of ownership of the Skibo estate some of the old, familiar family names which had long been inextricably tied to the history of Dornoch and Skibo—the Sutherlands, the Mackays, and the Grays.

From among these new owners, however, it was only William Gray who had funds sufficient to restore Skibo to what it had once been when the first bishops of Caithness had resided there. Gray eagerly pushed forward with the work begun by Mackay of restoring agricultural prosperity to the farmlands of Skibo. New farm houses were built and tenants were found for the deserted farms and crofts. The mills were put back into production and the small villages began to show renewed life. Gray also had plans for restoring the old castle and making it once again the palatial residence that Gilbert de Moravia had known.

Gray had the funds to carry out these plans, but unfortunately, he did not have the time. In 1760, after only four years of residence at Skibo, he died. His widow at first made an effort to continue her husband's work of restoration. Along with Dornoch Cathedral, Skibo Castle got a new slate roof, the first two such roofs in all of Sutherland. But Janet's trips to London became ever more frequent and more prolonged. She found life there far more congenial to her expensive and sophisticated social tastes than that which was available to her in remote and provincial Dornoch parish. Sometime in the early 1760s, she moved permanently to London, leaving Skibo to slip back once again into quiet desuetude.

In 1785, at the advanced age of ninety, Janet Sutherland Gray died in London. In the settlement of her estate, Skibo was again put up for sale and was quickly purchased by George Dempster of Edinburgh. Dempster had been born in Dundee in 1732, the grandson and namesake of a prosperous merchant and banker of that city. Young George had been educated at the University of Saint Andrew and had then received legal training at the University of Edinburgh, becoming a member of the Faculty of Advocates in 1775 at the young age of twenty-three.

These were the heady first years of the Scottish Renaissance when for a time Edinburgh became the major intellectual center of western Europe. With his wealth, education, and social standing as well as his intellect and quick wit, Dempster moved easily into that brilliant circle of philosophers and literati that included David Hume, William Robertson, James Boswell, Alexander Carlyle, and Adam Smith. He was admitted into Adam Ferguson's famous Poker Club, which later, with expanded membership and broader cultural scope, became the Select Society, a club which dominated Edinburgh's intellectual life. Within this glittering society, Dempster added his own shiny facet of high-level discourse on the philosophy of law and the pragmatism of practical politics.

In 1761, Dempster won a seat in Parliament for the burghs of Forfar and Fife by spending nearly £10,000 of his own money in a hotly contested election. For nearly thirty years, he represented these districts with such integrity and such independence of subservient party allegiance that he earned for himself a sobriquet unusual for a politician of that or any other age, that of "Honest" George. He vigorously supported Fox and Pitt in opposition to the Stamp Act and other colonial measures which, as he predicted would happen, led to the American Revolution.

Although living in London and Edinburgh and representing some of the larger urban areas of south-central Scotland, Dempster became increasingly interested in rural problems and the promotion of agriculture. Modern scientific agriculture was making its first tentative advancement in Britain in these decades with experimentation in the scientific breeding of cattle and sheep and with Charles "Turnip" Townshend's promotion of new food crops. Dempster took as his special field of interest the problem of soil drainage. He became convinced that much of Britain's marshy wastelands could be converted into arable fields, thereby not only making his country more self-sufficient in the production of food but also improving the health of its rural inhabitants by the draining of the miasmal swamps. He wrote numerous treatises on agricultural and fishing topics in these years, and his speeches in Parliament, his letters in agricultural magazines and his two major books, *Discourses . . . for Extending the Fisheries of Great Britain* and *General View of the Agriculture of the County of Angus and Forfar*, gave him an international reputation as a pioneer in scientific agriculture.

With the purchase of the Skibo estate in 1786, Dempster at last had the opportunity to put his theories into practice. In addition to Skibo, he was also able to purchase the neighboring estates of Pulrossie and Overskibo, this giving him a block of land of some thirty-five square miles in area, extending for fifteen miles along the north shore of Dornoch Firth and containing 22,000 acres.

Dempster arrived in Skibo in the summer of 1786 with his head full of grand plans for Skibo and his pockets full of wealth to make these plans a reality. Dempster's dreams for the future encompassed more than just agricultural reform, however. Before coming to Skibo, he had spent a few days at Mattock in Derbyshire where he visited Richard Arkwright's cotton mills. Why not, Dempster reasoned, bring the spinning industry to northern Scotland to revive its deserted villages and give employment to the displaced crofters of Sutherland? Far better that these uprooted rural folk remain in their old, familiar environment than push them southward to swell the slums of Glasgow and add to the misery of Edinburgh's narrow closes.

Dempster's first concern, however, must be with Skibo. Within a year, he had with a vengeance picked up the work of restoration begun by Mackay and briefly pursued by William Gray. Dempster's initial step in bringing modernity to Skibo was to resign the ancient feudal rights he had acquired along with the land. Believing that the best farmers would be those who knew that their tenure on the land was secure, he offered perpetual feu charters to those tenants he judged to be the most industrious and the most progressive in their farming methods. And of course, he immediately undertook the task of land drainage in those low lands that needed it. Skibo Castle was further renovated even though the family continued to live in the houses built by Mackay. The terraces below the castle were firmed up, and the gardens once again produced their bountiful crops of vegetables, fruits, and flowers that they had in the days of Gilbert's monks.

By 1790, Skibo was in such a prosperous shape that Dempster could turn his attention to his more grandiose plans for the whole region. In that year, assisted by his brother, Captain John Dempster, to whom he had earlier assigned the ownership of the Pulrossie estate, George Dempster began plans to build a new village which he named Spinningdale. Here he proposed to erect a cotton spinning mill and, wishing to make it the most up-to-date and efficient mill in Britain, he called in as a consultant Davide Dale, one of the country's pioneers in the development of the cotton spinning trade. Under Dale's supervision a large three-story cotton mill was erected, along with cottages for the hundred mill hands who would be needed. The mill had 36 jennies with 136 spindles for each jenny. The plant with its machinery was impressive, but unfortunately the output was not. Dornoch Firth was too remote from the main centers of the cotton trade in these prerailroad days. The cost of bringing the cotton bales by sea to far northeast Scotland and of shipping the finished cotton thread back to the main weaving centers of England proved exorbitant. Although the Dempsters and Dale continued to invest a great deal of money into the project, it remained a losing proposition. Finally, after a decade of effort on their part to make the enterprise profitable, a fire swept through the mill and the plant was closed forever. The ruins of the old mill still stand today, a picturesque souvenir of one of George Dempster's rare commercial failures.

A more realistic proposal was that which John Dempster

conceived, of building another village down on Newton Point just south of the castle, where the Firth narrows into its upper extension called the Kyle of Sutherland. Here Captain Dempster proposed to build a plant for the weaving of linen and the instruction of apprentices who could then take their learned weaving skills back to their individual cottages to give employment to the villagers and crofters. A warehouse was actually built down on the point in preparation for the mill, but the plans were never developed beyond that stage. So much money had to be poured into the Spinningdale venture that John Dempster did not have the funds available to pursue the Newton Point project further. Skibo was destined to remain rural and to miss the mixed blessing of Britain's industrial revolution.[2]

But George Dempster's agricultural revolution continued apace. Scientifically bred cattle and sheep improved the quality of the livestock for the entire area. Because he knew what he was looking for–and where to look–Dempster discovered large beds of marl in the region which could be used to fertilize the lime-deficient soil. Soon the farms of Skibo and those other farms using Skibo as a model were among the most productive and prosperous in all northern Scotland.

Always himself an avid sportsman in pursuit of the kingly salmon, Dempster next turned his attention to ways in which he could improve the lot of his neighbors who must fish for a livelihood. He taught them how to pack their freshly caught salmon in ice so that the fish might be sent to the rich and hungry markets in London. He was one of the promoters in forming a society for enlarging and protecting the fisheries of Scotland. This company built harbors, quays, and storehouses for the fish, but its good work was brought to an abrupt end when the war with France in 1793 disrupted the entire fishing industry.[3]

George Dempster's life was indeed a busy and profitable one. Here on his 22,000 acres of land, he had given dramatic evidence of what can be accomplished when swords are beaten into plowshares and spears into pruning hooks. Here at Skibo, after so many centuries of war, he had created a sanctuary of pastoral peace. It is probably safe to say that down the long line of Skibo's owners, from Gilbert de Moravia in 1225 to Andrew Carnegie in 1897, the only laird whose appreciation of and contributions to Skibo matched those of the saintly bishop and the great steel maker was George Dempster. Like his earliest predecessor and his later successor, Dempster found in Skibo a proper outlet for his love for the Scottish land and the opportunity to manifest that love by the creation of a rural idyll. It was Dempster who wrote of Skibo:

> It is the only habitable spot in Scotland I ever was possessed of. The sea, an arm of the sea, fine sailing, fishing, river and sea, dry and warm, rides on horseback—climate near the level of the sea, better than any I have as yet known. I wish it were still more sequestered, that one might once or twice in a season have a day to themselves in the country.[4]

These were lines to which either Bishop Gilbert or Andrew Carnegie might well have claimed authorship.

George Dempster had reason, in spite of the failures at Spinningdale and Newton Point, to be proud of his possessions and his accomplishments. In only one respect did he feel that his life was unfulfilled. He and his wife had no children to whom he could bequeath his beloved Skibo. When it was apparent that their union would, in the language of that day, remain without issue, George Dempster designated his brother's only son and George's namesake as his heir to Skibo. As John Dempster was often away at sea for months at a time, George Dempster became the child's guardian and father *de facto*. But little George was a sickly child and in spite of taking him south to winter in the milder climate of Exmouth, Devonshire, on the English Channel, and providing him with the best medical attention that money could buy, Dempster could not purchase health for his nephew and heir. The child died in Exmouth in April 1801.

One month later, Dempster received the belated word that the child's father and his only brother had gone down with his ship, the *Earl Talbot*, in the Indian Ocean the previous October. Quite suddenly, all of the beauty had departed from Skibo for George Dempster. In his deep despair, he wanted only to be rid of the place. Captain John Dempster had had one other child, born out of wedlock, a daughter who was much older than her legitimate half-brother, George. Harriet Milton was now a grown woman, married to a Lieutenant Colonel William Soper, and living with her husband in India. George Dempster wrote to his niece, whom he had rarely, if ever, seen, informing her that he was transferring the entire estate of Skibo to her. He and his wife then left Skibo to return to his father's old estate in Dunnichen.

The Sopers, with their small son, who was also named George, returned from India to claim Harriet's unexpected gift from her uncle. So that their benefactor's name would not be lost to Skibo, the Sopers added it to their own name, and thereafter were known as the Soper-Dempsters.

Upon transferring the estate to his niece, George Dempster had vowed that he could never again bear to see Skibo, but the old magic of this fairyland still had power to move him. He and his wife could not resist an occasional visit to Skibo, to become better acquainted with their newly acquired family and heirs and to observe how the land was prospering. The old magic of Skibo, however, still had the power to hurt as well as enchant. It was here at Skibo while on a visit to the Soper-Dempsters that George Dempster's wife died on 10 July 1810 and a few months later Harriet Soper-Dempster also died.

After these losses, George Dempster left Skibo forever. When he was not at his estate in Dunnichen, he spent long periods of time down in St. Andrews with his old friend Adam Ferguson, recalling the days of the Poker Club and the Select Company, when they were the lions of Edinburgh society and before George Dempster had ever heard of Skibo. Only once did he ever again return to the Firth of Dornoch. In 1815, he came back to dedicate a tablet in the toll house at Bonar Bridge commemorating the building of the bridge across the Kyle of Sutherland, a bridge which Dempster had actively promoted as a means of shortening land travel between Skibo and the main post road south of the firth that led to Inverness and on further south to the lowlands of Scotland. Three years later, on 13 February 1818, George Dempster died in Dunnichen, at the age of eighty-six, in what his friends could only regard as self-imposed exile from the land that he had loved best of all of his estates.

Lieutenant Colonel Soper-Dempster continued to reside at Skibo for many years after his wife's death. Upon his death, the estate passed to their only son, George, who stayed on at Skibo for much of the remainder of his life. Like his great-uncle, he too was childless. During their relatively long tenure of sixty-five years, from 1802 until George sold Skibo in 1866, the Soper-Dempsters seem to have left little record, either of a positive or negative nature, on the land and its tenants by which to mark their ownership. The farms were kept up, to be sure, and the tenants remained secure in the perpetual feu charters which George Dempster had granted them. But neither William Soper-Dempster nor his son George attempted any major improvements to the land or its buildings. The old castle keep once again deteriorated into a picturesque ruin, a suitable habitat for bats and the ghosts of ancient times.

In 1866, George Soper-Dempster, who was now in his late sixties, decided to sell Skibo. With no family but only old memories to tie him to this land, it had come time, he felt, to free himself of both the land and its memories and to live out his remaining years in a less isolated region. He found a purchaser in a Mr. Chirnside, who had but recently arrived in Scotland from his native land of Australia. Almost nothing else, including even his first name, seems to be known about Chirnside. Indeed, most chroniclers of the history of Skibo fail to mention that he ever existed and was for a brief time the proud—one assumes—Laird of Skibo. For what purpose he had come to Scotland, why he had purchased Skibo and then why after only six years, he sold the estate and hastened back to the "land down under" are questions for which the records provide us with no answers. As William Calder, the one historian of Skibo who does recognize Chirnside's existence, dryly comments, "There appears to be nothing remarkable about his [Chirnside's] personality, and on his departure he left 'no footprints on the sands of time' to distinguish his period of ownership."[5]

No one would ever so summarily dismiss his immediate successor, Evan Charles Sutherland-Walker, who purchased Skibo estate from the obscure Chirnside in 1872. This dapper, pugnacious little man came from a distinguished family of the Highlands, the grandson of Captain George Sackville Sutherland, laird of Uppat and also of Aberadar in Inverness-shire. Evan Sutherland-Walker appears to have been a curious amalgam of the traditional and the modern. In respect to dress and sought-after social status, he was a traditionalist, who took as his model the first Constable of Skibo, John Gray, the man who had first feued the lands of Skibo from Sutherland-Walker's remote ancestor, the Earl of Sutherland. Much to the amusement of the local inhabitants, Evan Sutherland-Walker almost always dressed in kilt, balmoral bonnet, and buckled shoes, while proudly displaying the tartan of the Sutherlands. How he revelled in the title, Laird of Skibo, and how bitterly he resented the fact that his predecessor, George Dempster, had so cavalierly dispensed with the ancient feudal rights that had once belonged to that title. He longed to return to an age when crofters were little more than serfs bound to the land and farm tenants were obsequious vassals. Like Edwin Arlington Robinson's Miniver Cheevy, Evan Sutherland-Walker "loved the days of old when swords were bright and steeds were prancing." He too "dreamed of Thebes and Camelot and Priam's neighbors" . . . and "cursed the commonplace and eyed a khaki suit with loathing; He missed the medieval grace / Of iron clothing."[6]

In other respects, however, Sutherland-Walker was a modernist who insisted upon the latest improvements in agricultural methods and who sought to make his gardens look as if they had been lifted directly out of the latest issue of *Stately Estates of England*. He built many new farm buildings at Skibo: stables, coachhouses, barns, and granaries. Sutherland-Walker also wanted the latest and most ostentatious style in his own living quarters. His romantic yearning for the past did not extend to the ancient castle keep, or to the graceful and simple Georgian houses which Mackay had built. His plans for the laird's manor house were more on the order of the late Victorian merchant princes of London, Birmingham, and New York, great piles of masonry that advertised their owner's wealth and social standing in every cornice and cupola. The old castle keep, or what remained of it, must go. Some of the older natives of the Skibo area today like to think that a part of Bishop Gilbert's castle still remains, incorporated into the massive structure that Sutherland-Walker erected, but that seems unlikely. It would have been more in character for him to raze to the ground what remained of that ancient historic structure and in its place build the castle of his dreams. What arose was an impressive semi-Gothic mansion of steeply pitched Flemish-style gables, ornate bay windows, and third-story cupolas. Initial impression was all-important to Sutherland-Walker, and no visitor to his home in passing through the vestibule could fail to be overwhelmed by the great baronial entry hall with its sweeping staircase at the rear leading to the floors above. Here before the large open fireplace in this great hall, little Evan Sutherland-Walker could sit in his high-backed carved-oak chair feeling every bit as regal as any Stewart monarch in his traditional royal Highland garb and knowing that he had the most currently fashionable residence that money and modern architectural skills could provide.

It was this extravagant quest for both modernity and traditionalism that finally undid Sutherland-Walker. Even though of distinguished and ancient lineage, he himself was not a man of great wealth. It had been necessary for him to borrow heavily in order to purchase Skibo, and that initial debt was multiplied many times over by his modernization of the farms and the building of his palatial residence. Modernity had plunged him deeply in debt.

But what finally destroyed him financially was his equally frantic effort to reverse the present economic and legal status of the estate and to bring feudalism back to Skibo. Fancying himself an expert in feudal law, Sutherland-Walker was determined to undo the land reforms of George Dempster by bringing suit against these tenants who for the past century had held the perpetual feu charters granted them by Dempster. Sutherland-Walker, to be sure, was motivated to take this action not only by his personal vanity of wanting to be a true feudal laird but also by the very pragmatic need for ready cash. He desperately needed to pay the architects, the contractors, and the suppliers of building material for his barns and mansion. To do so he must get more income from the estate. So off to court he went, confident in his claim that this was entailed land forever feued by the first Earl of Sutherland and by the bishops of Caithness under terms which could not be altered or revoked. George Dempster, he insisted, had acted illegally when he had unilaterally granted charters of perpetual feu. These charters must now be declared null and void.

But the Court of Sessions in Edinburgh in these last years

of the nineteenth century was not prepared to roll history back to the twelfth century. Sutherland-Walker lost his first case. Returning to Skibo in a fury, he gave orders to place a high fence around the victorious tenant's farm and to plant a dense, prickly hedge alongside the fence to box in the victorious tenant. Angry and frustrated but undaunted in losing his first case, Sutherland-Walker proceeded to take, one by one, each charter-holding tenant to court, only to lose each and every suit. To his former debts he was now obliged to add the staggeringly high legal costs of these fruitless court actions. In order to meet these new obligations, he turned his attention to those unlucky tenants who were not fortunate enough to hold Dempster's charters of perpetual feu. He raised the rental fees on these lands so high as to justify a new Peasants' Revolt. There was no revolt, but his tenants, unable to pay the increased rental fees, left the land. With no tenants, there was no production and no income.

In desperation, Sutherland-Walker drastically cut his own family's living expenses. He sold his fine carriages and horses, dismissed his liveried coachmen and grooms, and finally reduced his household staff to one or two servants. But it was to no avail. The tenants and servants were gone, but the creditors were still there unappeased. The only escape was through the ignominious door of bankruptcy. By court order, he was forced to leave Skibo. He sought temporary refuge at Pulrossie farm, which he had earlier deeded to his son. From there, after turning all that remained of his personal property over to the court-appointed trustees, he and his wife went south to Inverness. Out of respect for his family, the Estate Trustees granted him a small annuity which allowed him to rent a small cottage in Inverness. Here in his two or three rooms that would easily have fit within the confines of the great entry hall at Skibo, the Laird's grandiose dreams came to their pitiful end.

In the century and a half since the last battles were fought on her soil, Skibo may have been tamed and domesticated, but she still required expert handling. This was no place for the amateur. Pettifoggery and costume charades would not suffice. Skibo, as always, still demanded from her lairds love for her lands and respect for her people. She was soon to find a new laird who would give both in abundance and Skibo would know a greater glory than she had ever known before.

III

Skibo Finds a Worthy Laird

1897 – 1899

On a night early in April 1897, high upon the snow-capped summits of nine hills that tower above the Spey River valley in the parish of Laggan located in remote Badenoch of the central Highlands of Scotland, nine great bonfires were lighted. These nine fires, suddenly illuminating the cold night sky and giving temporary warmth to the men who had set them ablaze, must have caused fear in the startled red deer in the dark forests below and have aroused panic and much calling and flapping of wings of the wild fowl, nestled down for the cold night. But for the tenants of Cluny and the villagers of Garvamore, Crathie, and Laggan Bridge in the valley of Spey, it was a moment to cheer and a time to start the dancing, the playing of the pipes, and the drinking of toasts. They well knew what message this ancient Gaelic ritual of the lighting of the fires was conveying. A child and heir had been born to their laird. Word had finally come that the child had been safely delivered on the 30th of March far across the sea. So now the fires must be lighted, as such fires had always been lighted since the first Celtic Scots had come to this wild and beautiful mountain land two millennia before. With these fires, the tenants would celebrate a new life and give pledge of continued allegiance to an old order.

The infant whose birth they were celebrating on that cold night of April was not, to be sure, the child and heir of their true laird who owned Cluny Castle, the ancient seat of the Chiefs of the Clan Chatton. That title belonged to Colonel Ewen Macpherson. But he was an old bachelor, who gave no indication of providing them with an heir, and since inheriting the castle and its policies from his brother, Duncan Macpherson, Ewen was seldom at Cluny. For the past several summers he had leased his estate to an American millionaire, Andrew Carnegie, and it was for this American child that the fires were lighted. Most of the villagers sincerely hoped the Carnegies, whose occupancy of Cluny had brought unexpected prosperity and renewed fame to this historic spot, would soon purchase the estate from its absentee laird and thus ensure the continuation of the region's prosperity. And now with the almost miraculous birth of a child to their deputy laird, who was past sixty years of age, and to his wife, who had just celebrated her fortieth birthday, that hope of the parishioners of Laggan seemed a more real possibility than it ever had before.

In the imposing stone mansion on 51st Street on the island of Manhattan, three thousand miles away, much the same hope had already been expressed by the mother of the newly born infant. When Louise Carnegie, exhausted from her labor, first roused herself to greet her husband, who had come into her bedroom to embrace her and to meet his new daughter, she spoke of that which was most urgently on her mind now that their child had actually arrived, alive and well. As Carnegie would later remember their conversation at this very special moment, she said:

> "Here, Andrew, is your daughter. Her name is Margaret after your mother. Now one request I have to make."
>
> "What is it, Lou?"
>
> "We must get a summer home since this little one has

been given us. We cannot any longer rent one and be obliged to go in and go out at a certain date. It should be our home."

"Yes," I agreed.

"I make only one condition."

"What is that?" I asked.

"It must be in the Highlands of Scotland."[1]

That Louise should wish to make a second home in the Highlands hardly came as a surprise to her husband. She had been enamoured of Scotland ever since she had first seen it as a bride ten years before.

Carnegie had brought his wife directly to Britain following their marriage in New York on 22 April 1887. After a brief honeymoon spent at Bonchurch on the Isle of Wight, they had headed north to his native land and during that first summer of their marriage, they had leased Kilgraston House, a large country home just south of Perth and twenty-five miles north of Carnegie's birthplace in Dunfermline. Immediately upon her arrival north of the border, Louise became a Scotswoman. As her husband said in a speech he delivered that summer in Edinburgh, where he was given the Freedom of the City, "I begin to see in her the great danger that exists in all converts–she is beginning to out-Herod her husband in her love and devotion to Scotland. There is no tartan she sees that she does not want to wear. . . . Only today when she was up at the Castle of Edinburgh, she whispered to me, 'Oh, I know we want pipers at Kilgraston.' . . . Ladies and gentlemen, believe me that my wife is as thoroughly Scotch already as I am myself, and I leave to your imagination to say what she will become when she lives here as long as I have had the pleasure of living among you."[2]

Kilgraston had been delightfully beautiful during that first summer of their marriage, but it was not historically important enough to satisfy Carnegie's romantic love for Scotland's past, nor large enough to meet the needs of his constant entertaining. Louise had found that first summer somewhat difficult in being hostess to the numbers of visitors who flocked to their home. By nature a quiet, reserved, and private person, she had been unprepared for the public life that marriage to Andrew Carnegie meant. Occasionally, she could not forgo giving vent to her feeling of weariness and sense of inexperience in her frequent letters to her mother:

> Well, Mother darling, as some of the party have gone to church and others are in their rooms I have shut myself in my little sitting room to have a little chat with you . . .
>
> Well, our house is now full. Tomorrow we expect to sit seventeen at table . . . . I really have no actual care but it oppresses me to have so many people around. I see very little of them except at table and while we are driving but it all seems so sudden; there has been no growth, no gradual transition. I seem to be leading two lives—outwardly I am the mature married woman, while inwardly I am trying to reconcile the old and new life. I get awfully blue sometimes but I know it is very wrong to indulge in this feeling and above all to write it to you, but, Mother dear, I feel so much better for it . . . .[3]

And again:

> Well, Mother Mine, we are in the whirl, nothing but a rush and a bang all the while. I begin to experience the realities of life now and oh! how I do long for Mother!

But Louise was a proper young Victorian matron, carefully indoctrinated in the values and priorities of a nineteenth-century patriarchal society. She concluded this letter to her mother in the manner and style which that society expected of its women:

> I am not a bit homesick, but I begin to realize how much a man wants and how important it is for a woman not to have any wants or wishes of her own.[4]

And what Andrew wanted were crowds of people around him during all of these summer months of their honeymoon—Scottish relatives to whom he could show off his beautiful new bride, Scottish nobility whom he, the son of a poor immigrant weaver of Dunfermline, could now dazzle with the wealth of his possessions, and British intellectuals who could stimulate his active mind and insatiable intellectual curiosity and who, in turn, could be impressed with his erudition and worldly understanding. During this first summer, Louise wrote of entertaining and being entertained by the Gladstones, Robert Browning, Edwin Arnold, John Morley, the Yates Thompsons, William Black the novelist, and Lord and Lady Rosebery, as well as having the James G. Blaines, the Alexander Kings, the Courtlandt Palmers and Andrew's many Dunfermline relatives as house guests for extended periods of time. Obviously if this were to continue, and Carnegie gave every indication that it would—they must find a larger place for both their guests and their domestic staff. It was then that Carnegie found Cluny Castle and leased it from Cluny Ewen Macpherson for the following summer.

In late May 1888 they were back again in Britain for the second summer of their marriage. After a delightful three weeks of coaching through England with the Blaines—a trip which Louise particularly appreciated inasmuch as she had not been allowed to go on a similar coaching trip to which Andrew and his mother had invited her prior to their marriage—the Carnegies arrived at their new home at Cluny, which Louise had not previously seen. As their four-in-hand coach swept around the curve of the road and up the hill upon which the castle is situated, the coachman blew on his post horn and was answered by a salute from a cannon on the front lawn. The servants, an impressive assemblage, were standing by the front entry and waving white handkerchiefs of welcome. On the gravel drive in full regalia was John Macpherson, the son of old Angus Macpherson, Invernesshire's most celebrated piper, and himself a master piper, to pipe them home. For the young woman from Gramercy Park, New York, it was a moment of enchantment, a lovely fairy story in which the coach would never turn back into a pumpkin and the glass slipper never be lost.

"A whole week has passed in this delightful place and I never passed a happier one," she wrote to her mother. "We are all in love with Cluny already." And then she described in detail the wonders of this romantic Highland retreat.

> Such walks, such drives, such romantic little nooks! Imagine the most beautiful mountain brooks, one on each side of the Park, with rustic bridges, beautiful waterfalls, plenty of shade trees and shrubs and all surrounded by high rocky mountains with not a tree on them—nothing but rocks and heather—and you have some idea of it all. It looks in places just like the scenery in *Die Walküre*, and we are constantly pointing out where Brunhild is lying on the rocky summit surrounded by fire, or by the side of some beautiful brook the place where Siegfried comes upon the Rhine maidens, and then Walter Damrosch bursts out into song and sings passage after passage of the operas.[5]

This first week of enchantment continued on for a decade of summers, years which Louise would later remember as the golden years of her marriage, interrupted only by that fateful summer of 1892. This was the summer of the tragic Homestead Strike, which forever left its scar on the Carnegies, and the one summer in which Cluny Macpherson would not lease them his estate because he wanted to do some necessary repairs on the castle. The Carnegies had been forced to find an unsatisfactory substitute in a lodge on Loch Rannoch in an even more remote and isolated part of the Highlands. Louise and Andrew hoped they would never have to endure another season like that one, and Macpherson had reassured them that Cluny would be theirs to lease for as many summers as they wanted it.

As the years passed, Louise became the gracious hostess and competent manager that her husband wanted and demanded. Although there were still moments of weariness and times when she craved solitude, she knew how to direct the ever-present mob scenes of guests and servants, even how to deal with such embarrassing moments as that when they were entertaining the Gladstones and Gladstone's political arch-enemy, Joseph Chamberlain, dropped in unexpectedly for a call.

The crowds never diminished, but always seemed to increase, and when Andrew would inform her that still another party of friends was arriving and she would cry out, "We haven't the slightest room for them," his answer was not too helpful. "Oh, you can put them in my dressing room." The potential sleeping capacity of his dressing room soon became a family joke, but Louise managed to cope and even to enjoy this life which marriage to America's leading industrialist and one of the world's wealthiest men entailed. Andrew had promised her a rose garden and that he had most assuredly provided in profusion, but he had never promised her dullness and solitude. In some respects, the five months of summer which they spent in Scotland were the best months of the year for their marriage. For even though the crowds of guests were there, Andrew was also there. No extended business trips and social obligations to take him away from home as was the case when they were in New York. And there were those precious moments when they could be alone together, to walk through their gardens and to tramp through the heather on the high moors above Cluny. In one letter to her mother, she described those furtive trysts away from their omnipresent house guests:

> While this [Cluny] is a much finer place than Kilgraston, it is much more homelike and I just revel in it. From my little sitting room, steps lead right to the lawn, and Andrew's business room opens from my sitting room, so we slip from each other's rooms and out to the lawn with the greatest ease. Just now as I sit by the window writing, I saw a rabbit peep from under a clump of rhododendrons, jump around a little and run back. The lawn is like velvet and the flowers bloom continually. They have no end of gardeners about the place, and they keep it in beautiful order.[6]

And so the happy summers passed. Louise took up photography with great enthusiasm and captured on film for her family back in New York some of the beauty of the place which she so graphically delineated in her letters. Andrew had no interest in hunting, but he became an avid fisherman for the brown trout that swam the streams. Together, they took up horseback riding again, which brought back memories of their courtship days on the bridle paths of Central Park fifteen years earlier. It was an idyllic life and both were quite content to keep everything as it was.

In the fall of 1896, the Carnegies returned to New York with the happy knowledge that Louise was pregnant. Next summer when they came back to their Cluny, they would be returning with their child. But always there was the gnawing realization that Cluny was not really theirs—theirs to lease, to live in and love, but not theirs to have and to hold for their child. This kept worrying Louise all through the long winter of her pregnancy, and this had been uppermost in her mind when she first greeted her husband after Margaret's birth. So Andrew gave his pledge. They would have their own Highland home. He at once sent off a letter to Cluny Macpherson informing the laird that they wished to purchase his ancestral estate and would meet any reasonable price that Macpherson might set.

But it was not to be. While up in Greenwich, Connecticut, where he had gone for a short visit only to catch a severe cold which kept him in bed for several days, Carnegie received word from Scotland that there would be no sale of the Cluny estate. Carnegie wrote his wife from his sick bed:

> . . . Here is great news, a note from Mr. Macpherson of Cluny engaged to a Miss Hacey; funny name! Marriage not yet fixed . . . . Mr. Macpherson says nothing about future plans. *Nine* bonfires around Laggan in honor of Margaret; aren't they devoted? We must remain in that district, but hopes of our getting Cluny seem faint now, although there's many a slip, etc. . . . .[7]

There was, however, to be no slip. Carnegie was as mistaken about that as he had been about the name of Macpherson's fiancée. The impossible was about to happen. Old Cluny Macpherson, a bachelor of sixty-one years, was about to take as his bride a Miss Mary Stacey, eldest daughter of Cyril Stacey, a vicar in Gloucestershire. The summer of 1897, which would be Margaret's first summer in Scotland, would also be their last summer at Cluny. Carnegie must now spend as much of that summer as was necessary to find for his family their own Highland home.

Louise Carnegie brooded over this sad fact all that spring and upon arriving at Cluny in late May, it became increasingly hard to accept the loss of their home when she saw how her infant daughter was thriving in the bright sunshine and pure mountain air of Cluny. Louise was not as self-effacing and as submissive to a man's world as the letter written as a bride to her mother and quoted earlier would imply. She was a woman of courage and strong will, willing to go after what she truly wanted—and she truly wanted Cluny. Soon after arriving in Scotland, she decided that she herself would negotiate directly with the love-struck Cluny Macpherson. She wrote him a letter telling him that Cluny Castle had for the past nine summers become increasingly dear to both her husband and her and it was doubly so now that they had a child. If Louise hoped that a mother's appeal might succeed where a businessman's offer had failed, those hopes were quickly destroyed. Perhaps inspired by Carnegie's example of having sired a child at sixty-one, Cluny Macpherson, who was one year younger than his illustrious summer tenant, remained adamant. There would be no sale. Cluny Castle would remain in the Macpherson family, and the Carnegies must look elsewhere for their 'ain home.'

Carnegie had already begun the search. He had turned that matter over to his friend Hew Morrison, Librarian of the Public Library in Edinburgh which Carnegie had given to the city, and asked him to find them a suitable summer home. Carnegie made only three stipulations. Whatever place he purchased must have a view of the sea (which Cluny lacked), a trout stream, and a waterfall. Morrison was the right person to turn to for guidance for no one was better versed in the historic homes of Scotland or more *au courant* as to what might be available than the Librarian of Edinburgh.

Almost at once, Morrison arrived at Cluny with a suggestion and with the appropriate maps and papers to document that suggestion. There was an estate in Sutherland, right on the Firth of Dornoch, with a magnificent view of the sea, that was available. It was called Skibo—the ancient residential seat of the bishops of Caithness, later the home of the Gray family, George Dempster and most recently Ewen Sutherland-Walker. Few castles in Scotland had a more illustrious history than did Skibo. Morrison briefly outlined Sutherland-Walker's financial difficulties and told Carnegie that the estate trustees, who now held the property under the authority of the court, were eager to sell the estate and pay off the heavy debts that Sutherland-Walker had incurred. Skibo was a juicy plum, ripe for the picking at a fraction of its true value. As

Carnegie examined the various reports on the farms, crofts, and policies that made up the estate with that same sharp intensity and exquisite attention to detail that he always gave to the monthly reports of his steel mills in America, Morrison was forced to admit that most of the farms, buildings, the fields, and the roads were in deplorable condition due to Sutherland-Walker's bankrupt finances and to recent neglect. But, Morrison insisted, the potential was there. With money and care the estate could be restored to what it had been in George Dempster's day—one of the finest and most productive estates in all of Scotland. And the imposing mansion which Sutherland-Walker had built and had bankrupted himself in the building was still in excellent shape, ready for immediate occupancy with no renovation necessary.

Morrison gave Carnegie as hard a sell as the most forceful realtor could deliver. But Carnegie was not interested. Pushing the maps and papers aside, he told Morrison they would have to look elsewhere. Carnegie had already planned a coaching trip with a few friends through the Highlands, and he now suggested that Morrison accompany them on the tour. This would give them the opportunity to inspect various estates in northern Scotland that might be available. Morrison readily accepted this invitation. He had not dismissed the possibility of Skibo even if Carnegie had.

Early in June, Carnegie, Morrison, and friends set off on their two-week coaching trip, leaving Louise at Cluny with her infant daughter. The party inspected several estates that Morrison believed the Duke of Sutherland might be persuaded to sell, but none of them proved satisfactory to Carnegie. They were too far from the sea. Carnegie wanted a place easily accessible to his yacht. As the coach approached Bonar Bridge at the head of Dornoch Firth, Morrison, who had been quietly biding his time, again brought up the subject of Skibo. At Bonar Bridge, they would be only a few miles away from

Skibo. Why didn't he and Carnegie take a short side excursion, go down the road apiece and at least take a look at Skibo? Reluctantly, Carnegie agreed. Rather than take the coach and entire coaching party on what he was convinced was a fool's errand, Carnegie insisted that he and Morrison rent a wagonette in Bonar Bridge and go down themselves for a quick look.

As the two men and the driver bumped down the badly surfaced road that led to Dornoch, it was as if nature itself were conspiring with Skibo to give the old estate the laird it deserved. The day was perfect, as only a day in June can be in this region of Sutherland. The bright sun turned that narrow arm of the sea along which they rode into gleaming silver, and in the distance could be seen the thousands of acres of empty moorland—a hazy green now with only a slight lavender suggestion of the deep purple they would become when the heather was in full bloom. The wagon turned into the gate, down the long alley of ancient beech trees and yews that had been first planted by Gilbert's monks seven centuries earlier, and on into the circular drive before the white sandstone mansion that had been Sutherland-Walker's pride and joy—and ruin. There was no waterfall, but there was the sea, a perfect spot for a yacht, and the high hills of Carn Bhrain and Ben Tharsunn beyond. The birds were singing, the sun was shining, and the old magic still worked. Skibo was truly Schytherbolle after all, a fairyland of peace, and Carnegie was a quick and easy victim of its enchantment. Never mind the potholes in the road, the broken-down walls and fences, the tenantless farm fields growing rank with weeds. This was the place that Carnegie wanted.

Back at Cluny, the lonely and disconsolate Louise was hoping for the best from her husband's search, but fearing that nothing he found could ever match Cluny. It was, however, necessary that she encourage him to search diligently and select carefully their new home.

> . . . Am very anxious too for your report by word of mouth. . . . We now want to take root. We haven't time to make mistakes; as many playthings and play-places as you like and yachts galore, but a *home* first *please*, where we can have the greatest measure of health. . . . I'll try to be happy wherever you settle. We shall gang far ere we find anything muckle better than Cluny for baby. Her cheeks are as brown and as fat as possible. She almost talked to me as I was undressing her this evening.[8]

It is not surprising that Louise should be conscious of the passage of time. Andrew was nearly sixty-two, and she was past forty, but their daughter was only three months old. It was time to take root.

Carnegie returned home, exhilarated as he always was after a coaching trip, but this time, truly excited. He described for Louise the wonders of Skibo—the sunshine, the soft air tinged with the smell of salt from the sea, the rich farmlands lying fallow, and the magnificent grounds of the castle. Of course, if they kept the existing mansion, it would have to be greatly expanded. Not even the most capacious of dressing rooms in that place could accommodate all of the guests they would be having at Skibo.

Louise Carnegie was a bit overcome by Andrew's enthusiasm. Strong as was her desire to find a home of their own as quickly as possible, she now had to urge caution upon her husband, for as always, he was ready to rush forward headlong with quite undeliberate speed once he had decided to move. She pointed out that after all he had only seen the place for a few hours. Better to try the coat on first to see if it fit before purchasing it. Carnegie agreed. He wrote Morrison asking him if he would inquire of the Trustees as to the possibility of leasing Skibo for a year with first option to buy. With such a prospective purchaser in the wings, the estate trustees were only too happy to comply with this arrangement.

Late in the summer, Carnegie caught another bad cold which left him weak and exhausted. As it would be necessary to leave Cluny somewhat earlier this year than was their usual custom, the Carnegies decided to go south to London, where Carnegie could have expert medical attention before going back to New York. So early in September, the Carnegies departed from Cluny for the last time, and Louise's heart was heavy with sad farewells to the place and the people she loved. The crofters and cotters who waved them off also had heavy hearts. They had come to regard Andrew and Louise Carnegie as their Laird and Lady of Cluny. Now once again these people of the Spey valley were losing the opportunity of having a family in residence who would have ensured prosperity and fame for their communities. Fifty years earlier, Laggan parish had had the chance to provide the then young Queen Victoria with her summer place. Like Louise, Victoria in the 1840s on a visit to the region had been so taken with its beauty that she decided this was the place for a summer retreat for her and Prince Albert. But the place she chose had a tenant on a long-term lease. Undoubtedly she could have forced the issue and obtained the estate, but not wishing to achieve the same notoriety that the Duchess of Sutherland received in driving the tenants off the land, Victoria quite wisely looked elsewhere and found her castle at Balmoral. Now with the departure of the Carnegies, Laggan's dreams of wealth and glory once again faded away.

Later in the same month of September after the Carnegies' departure, Cluny's old laird married his Miss Stacey and came back to claim his ancestral home. But whatever dreams he may have had of continuing the Macpherson line at Cluny into the next generation also quickly faded. Ewen Macpherson died within three years of his marriage, and Cluny was passed on to still a younger brother, who was also childless. Soon after the turn of the century, the Cluny estate was put up for sale, but the Carnegies were not among the bidders. By that time, they were happily ensconced at Skibo, and Louise's heart was no longer in the Highlands of Badenoch watching the deer.

Arriving in London, Carnegie was attended by the best medical care that could be provided. These doctors warned him not to return to New York for the winter. Another severe cold in his weakened condition could lead to pneumonia, which would be very serious indeed. So the Carnegies rented the Villa Allerton in Cannes for the winter and left for the Riviera before the cold, heavy fogs of late autumn enveloped London.

From their villa that winter, Louise Carnegie wrote to her pastor in New York, the Reverend Charles H. Eaton:

> The giving up of Cluny, with all its tender associations, has affected my sister [Stella] and me deeply, and we cannot look forward to Skibo with the delight that Mr. Carnegie does, but when we have seen it and have lived there no doubt we shall grow to like it, particularly if it suits Mr. Carnegie and Margaret. We find our home here all we anticipated and more, and we are most delightfully situated. The restful out-of-door life has brought new vigor to Mr. Carnegie and really there is not a trace of his recent illness left. He climbs the hills with the greatest ease and is in the best of spirits.[9]

They were at Cannes in February when the United States battleship *Maine* was blown up in Santiago harbor in Cuba and in April when their country declared war on Spain. It was not a time to be crossing the Atlantic, and so any thoughts they might have had of returning to America before going up to their newly leased Skibo later in the summer were quickly dismissed. They stayed on at Cannes until Carnegie, always sensitive to summer heat, found it too warm to be pleasant. They then took a leisurely trip northward through France and England, accompanied by Louise's younger sister Stella, who since the death of their mother in 1890, had made her home with the Carnegies. It was not until early June that they reached Skibo. Carnegie wanted Louise to see Skibo at the same time of year that he had seen it. And again, Skibo put on its best face.

Skibo was not Cluny. It was not romantically wild, but gently pastoral. It did not have the scenery of *Die Walküre*, and one could not imagine here a rocky summit surrounded by fire upon which a Brunhild might lie. But there was the river, Evelix, so named by the ancient Celts because its rippling waters looked like beds of burning coal when set afire by the rays of the setting sun. And there was the sea, harnessed by the fairies' Gizzen Briggs into shimmering placidity. All of this Andrew excitedly pointed out to Louise, and his enthusiasm could not be other than contagious. Suddenly, there was no longer any doubt about the purchase. The coat had been tried on and quickly pronounced a good fit. Carnegie contacted his solicitor in Dunfirmline, John Ross, and told him to begin negotiations with the trustees for the purchase of Skibo. A price was soon agreed upon—£85,000 for Skibo and its 22,000 acres of land—a fair price, certainly, but for what Carnegie had in mind it proved to be only an initial down payment.

The summer of 1898, their first summer at Skibo, was unlike any other summer the Carnegies had spent or would spend together in Scotland. No streams of houseguests came that summer. Sutherland-Walker's mansion would simply not accommodate them. The Carnegies were far too busy with their plans for Skibo to have time for visitors anyway. So except for a short visit from Andrew's relatives in Dunfermline to inspect their illustrious American kinsman's latest acquisition, Andrew, Louise, and Margaret, with sister Stella quietly in the background, had the summer alone to enjoy each other's company. Louise was quite content.

Carnegie's first concern was to hire an architectual firm in Inverness to draw up plans for the elaborate additions that would be necessary to convert the existing mansion into the true Castle of Skibo that he envisioned. Next, he had to employ an able factor for the estate and a farm manager who together could find tenants for the vacant farms and crofts and could engage contractors and road builders to repair the roads and bridges on the estate. Being the successful executive that he was, Carnegie knew how to delegate. The actual details of planning and executing these plans were turned over to his designated lieutenants. Louise, working closely with the architects, would plan and design their home. The factor and the farm manager would restore the castle policies and the farmlands to prosperity and full beauty. Carnegie could then turn his attention to the internal affairs of his steel company and to the external affairs of his nation, both of which, he felt, were sadly in need of guidance. Carnegie had been out of the United States for fifteen months and although his steel company was earning greater profits than ever before, he needed to give constant reminder through frequent memoranda to his hard-working partners back in Pittsburgh that though far away he was still keeping an eagle eye on every detail of their operations.

As for his country, it too seemed prosperous and well. The Boy Orator of the Platte, William Jennings Bryan, and his wild, populistic idea for the unlimited coinage of silver had been, one hoped, forever buried under an avalanche of electoral votes in the presidential election of 1896. The benign William McKinley sat in the White House, gold was now king, God was in His Heaven, and all should have been well. But unfortunately, the devil was still at loose on earth doing his mischief in that summer of 1898. America was in the process of terminating its "splendid little war" with Spain after only four months of conflict, and as victor was about to collect its spoils. Carnegie's great fear that summer was that America was about to be spoiled by the very spoils which fell into her lap.

The Spanish American War had initially presented Carnegie with something of a dilemma. As an outspoken pacifist, he should have opposed that war. But as an equally vociferous opponent of Old World imperialism, he could not remain in opposition to a conflict which was concerned with monarchical Spain's last-ditch defense of the remaining bits of its once-vast American empire. So Carnegie had cast his pacificism aside and had enthusiastically supported his country's military venture to kick Spain out of the New World and make Puerto Rico as well as Cuba *libre*. But once nudged by America's not-very-powerful military arm, the entire Spanish empire quickly toppled and shattered. Now, to Carnegie's dismay, it would appear that the United States was about to pick up the pieces and glue them together into an American empire—not only Cuba and Puerto Rico, but also the vast Philippine archipelago halfway around the world. We would, like Britain, Spain, and Portugal before us, be holding subject peoples against their will under colonial domination. The American Republic would then, he was convinced, go the way of the Roman Republic, and Carnegie's much-heralded Triumphant Democracy would be replaced by Tyrannical Caesarism. It was time to yell "fire" from the housetops, and Carnegie spent most of that summer doing just that. He contributed heavily to Edward Atkinson's Anti-Imperialist League, he wrote numerous letters to his old friend John Hay, now Secretary of State, and even more letters to the editors of newspapers throughout Britain and the United States. His most important contribution to the anti-imperialist movement, however, was his essay, "Distant Possessions—The Parting of the Ways," which he wrote that August at Skibo and which was published in the *North American Review*. It was a strident alarm bell to awaken the United States to the dangers that overseas possessions had for Americans as well as Puerto Ricans, Filipinos, and Cubans, and Carnegie made sure that his articles got mass distribution throughout the United States.

All of these affairs kept Carnegie very busy, and it was left to Louise to instruct the architects as to the dimensions, form, and function of their additions to Skibo. First, it was decided to keep the central part of the mansion which Sutherland-Walker had built. Both Louise and Andrew liked the great front hall. Here under the stair rise and second-story gallery could be placed the pipe organ which they both wanted. The two front rooms off the hall which had served the Sutherland-Walkers as a drawing room and dining room could now become a morning room and small family dining room. The existing rooms to the rear of the great hall on the ground floor which were the kitchen and servants quarters would become space for serving rooms, pantries, and closets. A new wing would be built on the north of the existing mansion for a large modern kitchen, a tea room, a housekeeper's room, and staircase to the new servants' quarters on the upper floors of that wing.

The spectacularly grand part of the new Skibo, however, would arise like Venus from the shell of the old mansion on the south and west, facing the sea. On the southeast corner of this addition would be a large drawing room, a beautifully light room with floor length windows on the east and south, presenting to occupants of this room magnificent views of the terraces, gardens, and beyond, Dornoch Firth and the far-distant hills of Rosshire. Back of the drawing room would be the music room and, going further west along the corridor, one could enter the great library which Carnegie envisioned as being the very heart and soul of the castle. As Louise and the architects would design it, it became a glowing, golden heart indeed, panelled all in light oak with a pale yellow moulded ceiling. Above the twelve-foot-high built-in bookcases, with a capacity for seven thousand volumes, would run a two-foot-high frieze of carved oak, bearing the Coats of Arms of all of the Scottish cities that had given the Freedom of the City to Carnegie, bound together in the frieze by intricate Celtic designs of the fauna and flora of Scotland. The carved oaken fireplace mantel would bear a huge tome, carved out of a single block of golden oak, open to a page bearing one of Carnegie's favorite maxims:

> He that cannot reason is a fool
> He that will not is a bigot
> He that dare not is a slave.

Unfortunately, the library for all of the splendor of its appointments ultimately proved to be a disappointment. Carnegie had asked his good friend, Lord Acton, who probably had the finest personal library in Britain if not in the world, to select for him the six or seven thousand titles that Carnegie should have in his library. This Lord Action did and Carnegie was delighted with the selection. The ever-obliging Hew Morrison was then commissioned to find copies of each of the titles on Acton's list. Carnegie had specifically requested Morrison to find used—obviously used—copies of these books. "Please remember that I do not wish rare or curious books or elaborate bindings. It is to be a working library, only the gems of literature."[10] There was a kind of reverse snobbery in Carnegie's makeup that made him eschew the more obvious manifestations of wealth. J. P. Morgan's library of rare first editions had no appeal for him whatsoever. He wanted his library to look like Lord Acton's—a working scholar's library, not a rich man's hobbyhorse.

But Morrison, perhaps overawed by the magnificence of the room itself, could not really believe that Carnegie had meant what he said. Morrison obtained beautifully printed editions of each of the titles on the list, and then had them bound in the finest of gold, brown and green leather bindings. When Carnegie received the rather sizable bills from the bookbinders Morrison had procured, he let Morrison know his feelings in no uncertain terms:

The Dornoch Pipe Band visits Skibo. Summer 1980.

Skibo Castle. Detail. Southwest side.

Skibo Castle. Detail of tower with gargoyles. West side.

Skibo Castle. Aerial view of garden terraces.

> I asked you to get the best editions of a list of books Lord Acton would furnish you. I never said one word to you about changing the bindings of these gems, never. Now I learn that you have spent more money on bindings than the precious gems cost. This is, to my mind, not only a waste of money, which is wrong in itself, but an insult to the great Teachers from whom I draw my intellectual & emotional life—my spiritual existence. I would today give 3000 £ to have their messages cased in sober, cheap bindings showing it was *these* & not the bindings you have so strangely contracted for, that I value.
>
> It may be that I shall have to strip these Treasures of their costly covers & have their ordinary shells restored. I cannot tell until I see them at Skibo. I am really hurt by the affair & as I told you, I wish one check sent to you by my cashier (for I could never sign it) & charge the matter off once for all. It has to be paid of course for you were my agent.[11]

What Morrison may have said directly to Carnegie in response to this angry outburst is not a matter of record, but in the catalogue of the Skibo Library which Morrison prepared, he made the following statement of explanation:

> In the great majority of cases the books were already bound when bought, the binding being by men of eminence in their trade, as for example, Riverere & Son, of London, whose work adorns many of the examples in the class of modern literature. . . . The best binding material has been used, and where the leather was morocco, only the highest quality of French Levant was accepted. The gilding of the books (in accordance with the well-known wishes of their owner) has not been very elaborate, but it is in every case in keeping with the size, contents and importance of the volumes . . . .
>
> No rare copies of works or rare editions were sought

> after on account of variety alone. The only exception, it may be noted, is a copy of the first edition of Ruskin's Poems, as fresh as it came from the printer's hands, and this was added in order to make the collection of Mr. Ruskin's work complete.[12]

It is doubtful if this or any other statement could have satisfied Carnegie, but at least the bindings were kept on the books. Carnegie, however, never became reconciled enough to enjoy his magnificently bound treasures. They sat on their shelves, seldom if ever opened, as beautiful objets d'art to decorate the walls of the splendid room which had been built to house them.

Behind the library, in the southwestern corner of the new west wing that would be added to the mansion would be Carnegie's study, opening off the library. Here in this relatively small room, Carnegie would place his own personal working library—old, well-worn copies of books that he loved and made great use of. Next to this study was an even smaller office for his personal secretary, into which one entered from the corridor.

Beyond the entrance into the garden from this west wing would be the billiard room, and next to it, the large formal dining room, also panelled in light oak with a fireplace recessed in an alcove. On the walls of the dining room would eventually hang oil portraits of some of Carnegie's heroes, including among others, Benjamin Franklin, James Watt, and Robert Burns.

On the upper two floors of the new extensions would be eleven bedroom suites, five other more modest bedrooms, a play room, and fifteen bedrooms for servants. Most of the suites and bedrooms would bear place names of the region: Struie, Migdale, Ospis, Evelix, Dornoch, Laro, Lagain, and Cracail. But a few of the rooms would be named for certain illustrious figures who had played a part in Skibo's long history: St. Gilbert, of course; and Sigurd, the great Viking chieftain who reportedly lay buried in the earth of Skibo. The most imposing suite, however, aside from the Carnegies' own personal suites, would be named for the first Marquis of Montrose, that romantic and tragic defender of constitutional monarchy and religious toleration in the age of Cromwell, when religious bigotry and tyranny were the order of the day.

One of Skibo's proudest moments was that day in 1650 when Montrose, after having been defeated at the battle of Carbisdale near by, was brought to Skibo by his captors on their way to Edinburgh, where Montrose was to be hanged. The Laird of Skibo at this time was Robert Gray, one of the many of that name in the long line of Grays who possessed Skibo. Gray himself was at that moment a prisoner in Edinburgh, having been arrested for giving support to Montrose and the king. But his wife, the former Jane Seton, was very much in command at Skibo. When the ill-fated Marquis was brought under heavy guard to the castle door by Commander Holbourn of the Scottish Covenanter army, seeking lodging for two nights, Lady Jane Gray opened her castle to the welcome Montrose and his unwelcome captors. She gave orders to her staff to prepare a dinner for her illustrious guest. Since the castle was held only by a defenseless woman and her servants, Holbourn had undoubtedly come to Skibo because he considered it to be a safe resting place here in these dangerous Highlands where there were still so many covert supporters of the defeated Montrose. Surely this unarmed woman could give him no trouble. But Holbourn had not reckoned with the indomitable spirit of the chatelaine of Skibo. As the commander, his top lieutenants, and Montrose filed into the dining hall that evening, Lady Jane asked Montrose to sit at the head of the table in the place of the absent Laird. Holbourn was outraged. "Madam," he barked out in his most commanding military tone, "this man is a prisoner under sentence of death. He shall not sit at the head place." Holbourn then pushed Montrose rudely aside and he himself took the Laird's seat. Whereupon

Lady Jane picked up the leg of mutton already on the table ready for the carving, "Certes," she shouted, "if ye dinna ken ye're ain place an' manners at my table, I'll teach ye!" With these words, she hurled the leg of mutton down the length of the table. Her aim was straight and true. The roast struck the hapless Holbourn squarely on the chest, toppling him out of his chair and splattering his resplendent dress uniform with gravy and mutton fat. Picking himself up, Holbourn in high dudgeon stalked from the room. Lady Gray graciously motioned her guest of honor to his assigned seat, and then sweetly invited the remaining guards to seat themselves around the table. The company enjoyed a hearty meal, the roast being none the worse for its unusual wear and tear.[13]

Happily, Lady Gray's husband did not suffer the same fate of execution as did Montrose. Ultimately, Robert Gray received a pardon from Oliver Cromwell, but prior to being released from the Edinburgh prison, he had to pay a fine of 1200 marks in compensation to Commander Holbourn for the damage done to his uniform and the greater damage to his ego.

The Montrose story was a favorite of the Carnegies, who found in the Marquis's commitment to religious toleration and in the merciful amnesty which he always bestowed upon his own prisoners of war just reasons for placing Montrose high in the pantheon of Scottish heroes. Andrew Carnegie was particularly proud of the fact that Montrose's wife was Magdalen Carnegie, daughter of the first Earl of Southesk, a family to which he liked to claim remote kinship. So it was quite appropriate that the elegant, dark-panelled, fumed oak bedroom with its deep alcove containing one of the most elaborately carved fireplaces in the castle and with its great oriel window that looked out on the western hills would be named by the Carnegies "The Montrose Room" and would be reserved always for their most distinguished guests.

Thus did the plans take shape, room by room, during that first summer at Skibo in 1898. Because Louise was putting as much of her own creative thought and artistic sensibility into the new additions, Skibo in a matter of three months had become more personally hers than any other home she had ever known. It is not surprising then that the letter which she wrote to her minister, Charles Eaton, at the close of this busy summer was quite different in tone from the one she had written him from Cannes the previous winter when she was still grieving for the loss of Cluny and was full of doubts about Skibo:

> We are all very pleased with our new home. The surroundings are more of the English type than Scotch. The sweet pastoral scenery is perfect of its kind. A beautiful undulating park with cattle grazing, a stately avenue of fine old beeches, glimpses of the Dornoch Firth, about a mile away, all seen through the picturesque cluster of lime and beech trees. All make such a peaceful picture that already a restful home feeling has come. The Highland features to which our hearts turn longingly are not wanting, but are more distant.
>
> To show you the unique range of attractions, yesterday Mr. Carnegie was trout fishing on a wild moorland loch surrounded by heather while I took Margaret to the *sea* and she had her first experience of rolling upon the soft white sand and digging her little hands in it to her heart's content, while the blue waters of the ocean came rolling in at her feet and the salt breeze brought the roses to her cheeks. She is strong and hearty and so full of mischief—a perfect little sunbeam. With all our fullness of life before we have never really lived till now. . . .[14]

It had been good—this first summer at Skibo. When they left for New York in early October, they had the satisfaction of knowing that Skibo—their own home—would be waiting for them whenever they chose to come back. They were no longer tenants but master and mistress of their Highland home. Andrew could take pride in having selected and purchased Skibo, and he fairly purred with self-satisfaction over the reception which his anti-imperialist tracts had been accorded in the United States. Louise was equally content with the summer results—with the plans for the new Skibo which the architects had designed under her supervision and, above all, with how well baby Margaret had seemed to thrive in the sunshine and salt air of Skibo. But it was also good to be heading west toward America after eighteen months' absence—the longest Louise had ever been away from her native land.

Back in New York, there was no time to sit back in complacent contemplation of their past summer's achievements. Carnegie at once threw himself into the center of the great contest over American expansionism which the Spanish American war had engendered. The recently negotiated Treaty of Paris had given to the United States all that remained of the Spanish empire in the Western Hemisphere and the far reaches of the Pacific. That treaty must be rejected by the United States Senate, and Carnegie was prepared to use the power of his ample purse and what he liked to think was his even-more-powerful pen to save American democracy.

He was also coming to a crossroads in his business career, and there were critical decisions that must be made. The great era of railroad building was coming to an end. Carnegie Steel must look to new markets—structural steel beams for the skyscrapers that were beginning to change the skylines of Manhattan, Chicago, and St. Louis, and the possibility, perhaps even the necessity, of going into the manufacturing of finished products such as wire, tubing, and steel plates. But Carnegie was approaching sixty-five. Perhaps it was time to sell out to one of the vertically integrated steel trusts that were being formed at this time.

In spite of its size and the value of its annual product, the Carnegie company had remained a simple partnership. Carnegie had never even considered the possibility of forming a corporation with shares open for sale to the public. Within that partnership, Carnegie was the giant among pigmies, owning 58 percent of the total capital value. As Carnegie grew older, his partners grew more restless and concerned. For under the terms of their partnership, they were all bound by the so-called Iron Clad Agreement which stipulated that should any partner die, his share was not assignable to a designated heir but must be purchased by the remaining partners at the prevailing book value of those shares. If any of the other partners should die, even Henry Clay Frick with an 11 percent share, the partnership could easily make payment in cash to his heirs. But if Carnegie should die, rich as Carnegie Steel was, it would be bankrupted in paying off his share. Hence there was increasing pressure upon Carnegie from his associates to consider putting the entire company up for sale to some financial syndicate powerful enough to meet the purchase price.

Carnegie had previously resisted all such pressure from his partners, but now, to their surprised delight, he seemed more receptive. Perhaps Skibo had been instrumental in turning his interest away from the business of greater acquisition, now that he had acquired what he called his "heaven on earth." Perhaps he saw that if he was ever going to put into practice his own Gospel of Wealth, he had better start now. In any event, it was in this winter of 1898-99 that Carnegie gave permission to two of his senior partners, Frick and Henry Phipps, to see what they could do in regard to negotiations for the sale of the company. What they shortly came up with was a proposition from a syndicate of financiers who, they said, for the moment must be nameless, to purchase Carnegie Steel for $250,000,000. The syndicate asked only that it be given a ninety-day option to raise the necessary capital. Properly suspicious of financiers who couldn't be named, Carnegie demanded that for this option, the syndicate must deposit in trust two million dollars in cash of which $1,170,000 would be in his name, representing his 58 percent share of the company. Should the option to purchase not be taken up during the ninety-day period, the deposit would then be forfeited. With this guarantee of bargaining in good faith securely in his pocket, Carnegie began to turn his thoughts toward retirement and the dispensation of his projected huge fortune.

While Carnegie was thus preoccupied with his own empire of business and his country's business of empire, Louise found herself caught up in another major house-planning project. Some years earlier, Carnegie had bought several city lots far up Fifth Avenue on 91st Street, twenty blocks north of the last buildings of any consequence on that famed avenue. Most of New York's wealthy Four Hundred considered him to be crazy, for this was Manhattan's dumping ground for some of its unwanted population, refuse who had drifted northward to find a squatters' haven. Here these down and outers had built their lean-to-shacks and grazed their goats, unbothered by law, unconcerned about order. But Carnegie had foreseen the day when Manhattan's ever-expanding commercialism would swallow up the chateaux of the Vanderbilts and the Medici Romanesque palaces of the Goulds and his own more modest home on 51st Street. Now that he had tasted the joy of castle building in Scotland, Carnegie felt it was time for mansion building in Manhattan. And once again, after employing the distinguished architectural firm of Babb, Cook and Willard to draw up the plans, he turned the details over to Louise. It was as if she had never left Skibo. Once again she was faced with specification sheets, blueprints, and architectural drawings. But by now she felt herself almost a professional. There would be no consultations with Andrew. The entire house would come as a complete and finished surprise to her husband. A lovely retirement gift from her to him.

In late April 1899, they left New York for Britain, Andrew's head filled with politics and philanthropy, Louise's head with the building plans she had left behind in New York and the building plans that still awaited her in Skibo. After a leisurely month spent in southern England and London, the Carnegies arrived in Sutherland. As they drove in their open carriage down the old and now familiar road from Bonar Bridge to Skibo, all of their tenants and the people of the small villages were out along the road in their Sunday best, waving American and British flags to welcome their new laird and his lady. Carnegie was overwhelmed with emotion. At the entrance to Skibo, he stopped the carriage to receive an address of welcome from a ninety-year-old man, his oldest tenant. In response, Carnegie pointed to Mrs. Carnegie and said, "Here is an American who loves Scotland," and then pointing to himself, "and here is a Scotchman who loves America, and here," pointing to the two-year-old Margaret, "is a little Scottish-American who is born of both and will love both; she has come to enter the fairyland of childhood among you."[15]

Three weeks later, the builders were ready to lay the cornerstone for the great new southern addition to the castle. Carnegie insisted that his two-year-old daughter should give the final pat of the silver trowel on the top of the pink-white sandstone foundation block. Inscribed on the stone were the words, "Foundation Stone of the New Part of Skibo Castle Built by Andrew and Louise W. Carnegie Laid by Margaret Carnegie 23rd June 1899."

On the same day of this happy occasion, Carnegie wrote a letter to his cousin and business partner, George Lauder, to tell of the visit he had had from their other two senior partners, Frick and Phipps, the previous week. These busy negotiators had hurried across the Atlantic to be two of the first houseguests at Skibo in order to ask that the syndicate be given an extension of time in order to raise the necessary capital. "I said not one hour," Carnegie wrote his cousin. "I said business was to be so fine & next year would show [profits] of 40 to 50 m. . . . Meanwhile nothing to be said to our people except that option ceases Aug. 4th & they needn't be thinking over anything but attention to business &c."[16]

Carnegie would not be retiring after all—not just yet. And it was a most appropriate day—this day of laying the cornerstone—for him to be reporting to Lauder on the failure of Frick's and Phipps's syndicate to take over Carnegie Steel. For, much to their surprise and dismay, Carnegie insisted on keeping $1,170,000 deposit money on the option to purchase. By happy coincidence, that came almost exactly to the dollar the total cost of Skibo, including the building of the extensions to the castle. In later years, after his relationship with Frick was broken irreparably, Carnegie would delight in telling visitors, awed by the crenellated splendor of Skibo, "The whole thing is just a nice little present from Mr. Frick."[17]

Skibo did, indeed, have a new laird that was worthy of its grandeur. A laird that was not only a rich American but a canny Scotsman as well. Perhaps he was as rich as he was *because* he was such a canny Scotsman. At least, Scotland liked to think that was the case.

IV

Skibo's Gloria in Excelsis: *The Andrew Carnegie Years* 1900 – 1919

The Carnegies, on 16 March 1901, departed from New York for Europe for their fourteenth annual spring migration since their marriage in 1887. They left New York that year somewhat earlier than was their custom as they intended to spend a six-week holiday in Antibes and at the spa in Aix-en-Provence before heading north to Skibo. Carnegie had written his cousin George (Dod) Lauder, just prior to their departure, "I feel better get Louise away from this new house–architects &c too troublesome especially on Furniture & Decorations . . . getting six weeks holiday for her free from House & Housekeeping."[1]

Carnegie, too, was eager to be off, to be free of the constant badgering of newspaper reporters to which he had been subjected for the past month. For Carnegie had just sold his vast steel empire to a syndicate, headed by J. P. Morgan, which proposed to consolidate Carnegie Steel with Federal Steel and the various companies manufacturing finished steel products organized by the Moore brothers' syndicate. America's first billion-dollar corporation, the United States Steel Company, with Morgan officiating as midwife, was about to be born, and the presses on both sides of the Atlantic were loudly proclaiming that Carnegie's passing from the world's industrial scene marked the end of an era.

The end had come dramatically with a great flourish of trumpets and drums. Now Carnegie was strutting off the stage of business with what sounded more like a coronation march than the solemn final cadence of a Nunc Dimitis. He was leaving at the moment of his greatest power when his very success had threatened the ordered stability the finance capitalists of America were trying to achieve within the steel industry. His proposed new plant to manufacture finished steel products at Conneaut, Ohio, promised to be the largest and most efficient in the world, one which would give Carnegie Steel true verticality and which could easily destroy Morgan's National Tube Company and the Moores' various American companies producing wire, nails and hoops. But Carnegie's imperialistic plans for the new century were not limited to aggression within his own steel industry. He also whirled upon his old enemy, the Pennsylvania Railroad, under whose monopolistic control of rail transportation in western Pennsylvania he had long chafed. Carnegie had recently begun serious discussions with George Gould to put together a truly transcontinental rail line running from Baltimore through Pittsburgh and St. Louis and on to the west coast. It was as if Carnegie were beginning his business career all over again, but now with a base of power that could make him the supreme master of the American economy.

J. P. Morgan, sitting in his inner sanctum at 23 Wall Street, had watched these developments with a keen and most critical eye. Morgan, like John D. Rockefeller, was an orderly, systematic man. He did not really believe in the free enterprise system. He hated the waste, duplication, and clutter of unrestricted competition. And he had the power, the imagination, and the ability to build combines and trusts and interlocking directorates that could eliminate that wasteful competitive system. Now this maverick, Carnegie, whom his own industry had never been able to control through pooling arrangements and gentlemen's agreements, was out to destroy Morgan's world. Morgan never gave press interviews, but under these circumstances he felt compelled to utter one of his rare oracular public statements: "Carnegie is going to demoralize railroads just as he demoralized steel."[2] This was an ultimatum. Carnegie must be stopped.

Morgan was realist enough to know that there was only one way to stop him. This wild bronco could never be lassoed and brought into a syndicate corral as Morgan had so easily done with other free-wheeling industrialists of the open range. Nor could he be broken by competition. Carnegie had the resources in raw materials, plants, and technical competency that made him invulnerable to any combination of steel industries that might try that. There was only one way to meet this threat. A pasture so lush and green must be offered to him that Carnegie would be corralled by his own greedy appetite.

So working through Carnegie's vice-lord within his steel empire, Charles Schwab, whose own ambition led him to envision himself in command of a new super corporation, Morgan let it be known that he would meet any price Carnegie might set for his company. It must not have been an easy decision for Carnegie to make. He had never enjoyed empire building more than at this moment when the entire world of steel was within his grasp. But he was also acutely aware of the fact that he was now sixty-five. If he were ever going to start practicing his Gospel of Wealth which he so loudly preached, he had better climb out of the pulpit and start ministering to society. Above all, he knew what Louise's wishes were in this matter. It took only a night of thought, and then Carnegie had written with his usual blunt pencil on a scrap of paper a short note to Morgan, which he gave to the eager Schwab to deliver. He would sell all of his steel holdings for $480 million dollars. His own share of $270 million must, however, be in 5 percent, gold, first-mortgage bonds.

Carnegie had no intention of becoming the major stockholder in Morgan's new corporation, and he wished to have his payment in easily negotiable bonds to facilitate the disposal of them that he anticipated making.

It was as simple as that. Morgan accepted the price Carnegie had scribbled on the scrap of paper. The biggest sale in American industrial history had been consummated with all the formality of an errand boy's taking a shopping list to the corner grocery. When Morgan came up to Carnegie's home to consummate the deal with a handshake, he said, "Mr. Carnegie, I want to congratulate you on being the richest man in the world."[3] This was probably not true, but certainly no other person in the world had a fortune as large in liquid assets which could immediately be converted into cash.

Carnegie never wanted to see or touch a single one of these bonds that represented the fruition of his business career. The bonds were delivered to the Hudson Trust Company, Hoboken, New Jersey, to be stored in a special vault of impressive size. It was as if Carnegie, being the Celt that he was, feared that if he looked upon them they might vanish, like the gossamer gold of the Leprechaun. But with satisfaction, he wrote to his dear friend, John Morley, "I'll have *at least* 50 million [£] stg. all in 5% gold bonds & safe as any–and then. Ah, then, well, I'll tackle it. You'll see–I could as well had 100 million [£] stg. in a few years, but no sir, I'm not going to grow old piling up, but in distributing."[4]

The Carnegies embarked on the German liner, *Kaiseren Theresa,* in mid-March, a contented couple. It was a new century and they were sailing into a new world to which they looked forward with happy anticipation. For Louise, much of her contentment was attributable to the successful termination of her husband's business career, which she had never shared and had never liked. No more Homestead strikes and no more acrimonious disputes that had divided Andrew from so many of his old partners, particularly Henry Clay Frick.

Instead of the frenetic pursuit of ever-greater wealth, which she had never understood nor appreciated, there would now be the wise and humane distribution of that wealth back to society from whence it had come—a task which she did appreciate and in which she could participate with her husband. Her own burdens of supervising the building of two great mansions would also soon be laid down. Within the year she expected both the renovation of Skibo and the completion of the mansion at 2 East 91st Street to be accomplished. She and Andrew could then enjoy the products of their joint labor.

Carnegie also looked ahead eagerly to his new life as giver instead of getter. He prided himself on being the pioneer scientific philanthropist. He was not the first man certainly to give away the bulk of his great fortune—he always paid high tribute to Enoch Pratt, Peter Cooper, and George Peabody for earlier having done that with their wealth—but he felt justified in claiming to be the first to analyze the problem of the administration and distribution of great wealth and to state specifically and precisely "the best fields of philanthropy." He would shed light through his three thousand community libraries; he would advance knowledge through his great institutions in Washington and Pittsburgh and his university trust in Scotland; he would so enrich the town of Dunfermline, his birthplace, as to make it a model for humane and gracious living; he would bestow medals and pensions not upon military heroes who took away life, but upon those civilian heroes who saved life at the risk of their own lives. Above all, he with his millions would purchase peace for the entire world. He would build a great Peace Palace at the Hague to house a World Court which would settle by law rather than by guns all international disputes. He would endow a great trust to study the causes of past wars and by so doing find the appropriate means for the prevention of future wars. And when peace on this planet had been permanently established, there would still be funds left, he felt confident, to deal with the many other, lesser scourges of mankind. Carnegie had a drawer in his desk labeled "Gratitude and Sweet Words" in which he filed the letters he received from those communities and institutions that had already benefited from his largesse. He now happily anticipated within a short time that drawer would have several more companion drawers so labeled and stuffed with even sweeter words from a grateful humanity. All of this could be savored, reflected upon, and brought to pass as he luxuriated in the glories of Skibo, his "heaven on earth."

Upon their arrival back in Skibo in late May 1901, the Carnegies found that there still remained a great deal of work to be done before Skibo attained that state of perfection they had planned for their grand old estate. The magnificent new wings they had designed for the south, west, and north of the mansion were complete. Now they had to be decorated and furnished. The two eighteenth-century houses south of the castle which George Mackay had built a century and a half earlier and which Sutherland-Walker had allowed to fall into ruin were now torn down, but the land these houses had occupied must be terraced for walks, flowerbeds, and a fountain.

There were also the even more ambitious projects on the policies of Skibo that were well under way but far from finished: the new lake that Carnegie had ordered to be created to the west of the castle, Loch Ospisdale, to be stocked with brown trout which one could fish for from the bank; the smaller ponds, Lake Louise and Margaret's Loch, for water lilies and other aquatic plants; the building of a dam at the mouth of the Evelix River to create Loch Evelix with a salmon ladder so that the salmon could still go up the Evelix to spawn and to be caught; the nine-hole course so that Andrew could continue to take the medicine from what he believed to be the best physician in the world—Dr. Golf; and Carnegie's special pride and joy, a great, enclosed swimming pool, twenty-four meters long by nine meters wide, filled with salt water brought in from the sea and heated by an adjacent boiler and pump house. There were new farm buildings yet to be erected—dairy houses, barns, coach houses, and cottages for the farm workers. New lodge houses were also being built on the east and at what had previously been the entrance to the castle grounds off the Bonar Bridge-Dornoch road on the north. But the main entrance to the castle now would be on the west, through high iron gates, past an imposing Scottish baronial-style lodge house where the Chief Forester would reside with his wife who would serve as gatekeeper. The mile drive from the west gate would take the visitor past Pulrossie farm and the new Loch Ospisdale through an avenue of Scotch pines up to the castle door.

Carnegie's golden wand would bring about all of these amenities within the coming year. In the meantime, he was providing employment for much of the skilled and unskilled labor throughout Sutherland and Ross counties. The region had never before known such prosperity, and the sun shone down brightly upon this fortunate land. Skibo would give substance to John Keats's poetic conceit that "A thing of beauty is a joy for ever."

In a project as ambitious as this one was of converting an ancient and run-down estate into a magical fairyland of beauty, mistakes would of course be made, but they could be rectified with imagination—and a seemingly unlimited supply of funds. For example, in building the boiler house that would generate heat for the swimming pool and the greenhouses, the most modern and efficient steam plant was constructed equipped with a high stack to carry off the coal smoke from the furnaces. But the first time the boilers were fired, the engineers discovered to their consternation that they had located the plant in precisely the wrong place. The prevailing winds from the southwest carried the black smoke directly up the hill to the castle itself. There was however, an easy—if somewhat expensive–solution. A long tunnel was dug from the boiler house to the castle, and the offending smoke was then blown through the underground conduit into the furnace of the castle where it could escape out of the flue of the castle's main chimney.

The golden wand continued to wave, and in every direction that it pointed new splendors arose. No detail was too insignificant to be overlooked or disregarded. A local Clashmore man and his son were employed as "walkers," to walk the several miles of road on the estate in order to detect and remedy the slightest bump or rut that might possibly jar the family's carriages and carts in the smooth, if not "swift completion of their appointed rounds." Although this particular nicety of travel at Skibo ended with World War I, those men and their descendants even today still bear the sobriquet of Walker, although there are few local residents who remember the origin of the name.

In early October 1901 when the Carnegies left Skibo to return to New York they had the satisfaction of knowing that when they returned the following May, Skibo would be complete and perfect. Not a single bit of wood or stone carving still to be sculptured, nor another drapery or picture yet to be hung. The great organ would be installed in the reception hall, and on the first landing of the grand staircase would be the five leaded, stained-glass windows in which a brief résumé of Skibo's and Carnegie's own history would be captured in the brilliant color of stained glass worthy of a French cathedral: St. Gilbert of Dornoch, mitred head-dress in one hand and shepherd's crook in the other, quite properly in the center panel, with the date 1235; on his left the portrait of the noble Marquis of Montrose in armor and the date 1650; on the right, Sigurd, the Viking chief in full battle regalia, carrying an axe and dated 946; at the far right, the weaver's cottage in Dunfermline, Carnegie's birthplace, and the small sailing vessel, *Wiscasset,* in which he and his parents sailed to

America, dated 1848; and in the panel on the extreme left, the great ocean liner in which the Carnegies sailed back to Scotland and above that the majestic splendor of the new Skibo castle, dated 1898. Skibo had been witness to a great deal of history over the millennium of its existence, but no period had been more dramatic—or ultimately of greater significance to Skibo itself—than that relatively short span of fifty years that lay between the twelve-year-old Andrew's discovery of America in 1848 and his finding of Skibo in 1898.

The summer of 1902 marked the real beginning of Andrew Carnegie's lairdship of Skibo. Nothing more need be done to perfect the Carnegies' Highland home.[5] The next thirteen years were probably the happiest of Andrew's life. As he approached and passed seventy, he never felt healthier or more vigorous. The years seemed to have no effect upon either his energy or his capacity to enjoy life. And most of all, he enjoyed Skibo. The five months spent there were for him the very elixir of living. "I am so busy working at fun!" he wrote Dod Lauder. "Fishing, yachting, golfing. Skibo never so delightful; all so quiet. A home at last."[6]

Skibo may never have been so delightful, but Louise would have called it anything but quiet. Its remote location did not mean isolation, for here, far more than in New York, the Carnegies carried on an intensely active social life. Carnegie delighted in mixing his guests with a complete disregard for social background, politics, religion, and nationality. Old Dunfermline neighbors and cousins came face-to-face with prime ministers, poets, university professors, and American businessmen—it was good for all of them, Carnegie reasoned. What did matter was to have the house full. He thrived in great crowds and was delighted to see them come and was reluctant to see them go. For those guests of whom he was particularly fond he would, upon their departure, have raised over the main gate a large sign which could be spotted a quarter of a mile away: "Will ye no come back again"—a gracious but quite unnecessary gesture of farewell, for they would and always did many times.

Skibo was an exciting place to be but every guest no matter how distinguished his rank or eccentric his personal habits, had to adapt himself to the regimen that prevailed there. If a guest had any notion of sleeping later than eight o'clock in the morning, that expectation was rudely dispelled on the first morning at the castle. Promptly at that hour there would come a distant wail that to the uninitiated sounded like the crying of a lost soul in one of Dante's more deeply depressed circles in Hell. The by now fully awakened guest, looking out his bedroom window, would see a Scottish piper in full Highland dress, approaching the castle, skirling the pipes with all of the vigor of a piper leading the troops into battle at Culloden. It was Angus Macpherson, brother of John Macpherson of Laggan Bridge, who had been the Carnegies' first piper at Cluny.[7]

By now, sleep being out of the question, the hastily dressed guest would descend the grand staircase and walk, a bit self-consciously, through the great hall and down the long stone-pillared corridor to the dining room, again to the sound of music, more recognizable as such to non-Celtic ears, which emanated from the organ in the hall. The kippers, porridge, and scrambled eggs were eaten to the majestic tones of Haydn, Bach, and even Wagner, which echoed throughout the castle. It was an experience that a visitor to Skibo would find difficult to ignore while there or to forget after leaving.

Following breakfast, the guest had a variety of diversions from which to choose—trout and salmon fishing in the lochs, stocked from Skibo's own fish hatchery; swimming in the salt water pool; playing golf and, with luck, losing to Carnegie; or, if it was the season, hunting grouse, pheasant, or deer. In this last pursuit, however, the guest had to dispense with the company of the host, for Carnegie hated guns and would never take part in a hunt, which he regarded as not a sport but a slaughter. For the guest of more sedentary habits there was the magnificent library, but this apparently was a seldom-chosen diversion, and the magnificently-bound books remained in their pristine splendor on the shelves, studiously ignored by Carnegie and untouched by his preoccupied guests.

Although Carnegie was lavish and unpredictable in his issuance of invitations to both Americans and Britons to visit him at Skibo, in time there came to be an established pattern to the entertainment. The first week in September was known as "The Principals' Week." There might be other guests present at the same time, but this week was primarily reserved for the visit of the heads of the four Scottish Universities, who had met each other collectively for the first time at Skibo soon after the Scottish Universities Trust had been established in 1901. They had found the experience so rewarding that Carnegie continued it each year. Another week was reserved for the board of trustees of the Dunfermline Trust. Led by Sir John Ross, these solid Fifeshire men made their annual pilgrimage northward, and Carnegie found in their natural lack of affectation and in their impassive acceptance of the splendors of Skibo a rather delightful contrast with many of his other guests.

The Fourth of July was elaborately celebrated as Fête Day. All of the schoolchildren for miles around would gather on the lawns of Skibo to run races, drink lemonade, sing songs, and salute that curiously hybrid ensign, consisting of the Union Jack and the Star Spangled Banner sewn together, that fluttered from the mast of the highest turret at Skibo just as it had earlier flown from a flagpole at Cluny. Carnegie took particular pleasure in celebrating this day of American Independence on British soil, and he would caper about with all of the exuberance of the most youthful guests.

Then there was the group familiarly known to the Carnegies as the "Old Shoes," who were always welcome at Skibo and needed no special occasion for a visit. These were Carnegie's oldest British friends, most of them Liberal party leaders, who had first introduced Carnegie into British journalistic and political circles in the 1880s: the Yates Thompsons (Mrs. Thompson was Louise's closest friend), Lord Armistead, Frederic Harrison, Sir Henry Fowler, Herbert Gladstone, and Swire Smith. A few Americans were admitted into this select company on their visits to Europe: Richard Gilder, Andrew White, and Nicholas Murray Butler.

The "Oldest Shoe" of all was, of course, John Morley. With each passing year, Carnegie and Morley seemed to come closer together, each more dependent upon the other's friendship. This deep attachment between two men so radically different in background and temperament was difficult to understand, even by Carnegie and Morley themselves, but its inexplicability made it even more real and unexpectedly precious to them both. For each, the other served as the one person with whom any idea, any personal problem, any hope or despair could be shared. Their exchange of letters became more and more frequent, and their increasing affinity was reflected in the changing salutations of the letters over the years, beginning with that proper English informal formality of "My dear Morley," and "My dear Carnegie," progressing to "Dear Friend Morley" and "Dear Friend Carnegie," and finally reaching the point where the reserved Morley could write, "My dearest and oldest friend," and the much more effusive Carnegie would respond, "My Chum–Dear Chum." (Who else in all the world would think of referring to–much less dare to address–Morley as "Chum"?) Hardly a week passed during the summer months of each year that Carnegie did not write Morley, urging him to drop whatever political or social obligations he might have and come to Skibo. No matter how many others might be there, there was always a place for Morley. "Don't fail to come soon as you can next week," was a typical Carnegie letter. ". . . I hope you are to be greatly benefitted by your stay with us which must not be short—stretch it as much as possible. Bryce's may be here ere you go. Montrose Room will be ready for its rightful occupant & I'll be happy."[8] Morley would demur at being assigned the most luxurious suite at Skibo. "Now 'Montrose' besides being beyond my merits has one drawback. In these days I awake too early and want to read in bed. Montrose, if I am right, does not favour this practice. If it made no difference, I'd as lief have a room where the window shines upon the pillows of the guest. Excuse me for this petition if you please. . . . I am looking forward to it all with infinite satisfaction."[9]

Morley definitely belonged to the more sedentary group of Skibo visitors who spent the day trying to avoid all of Carnegie's planned fun and games. But only Morley, with his privileged status, could totally escape all of this vigorous activity. Carnegie quite early gave up an attempt to make an athlete out of him and was quite content to have Morley restrict his physical activity to taking long, slow walks with him along Sunset Walk, a trail that led from Skibo to Loch Ospisdale, where in the late evenings one could get a magnificent view of the setting sun over the distant moors and firth. Here on this walk or seated on stone benches on the castle terrace, the two men would discuss current politics, argue over literature, and laugh at ancient jokes. They delighted in baiting each other, for each knew the other would be quick to respond. For Carnegie these moments were quite the best times at Skibo.

The Carnegies frequently had unexpected visitors at Skibo, old friends whom Carnegie, in his usual expansive if indefinite manner, had urged to "come to Skibo sometime to see us." There seemed to be always room for more plates at the dining table or more cups on the tea cart, even more beds without having to utilize Andrew's dressing room. Louise was expected and always was able to cope. But one such uninvited and unexpected guest did cause something of a stir. One warm, sunny afternoon in July 1903, when the Carnegies had retired to their rooms for a much-needed nap, Louise was aroused by the housekeeper with a message from the Duke of Sutherland in residence at his castle in Dunrobin. The telegram informed her that King Edward VII was on his way from Dunrobin to pay a call on the Carnegies. Buckingham Palace was in the process of being renovated and the King had heard so much about the marvels that the Carnegies had effected at Skibo that he wished to view them firsthand. No time for elaborate preparations—the King would be there within the hour and would have to be served the same ordinary tea as if he were Cousin Annie Lauder making an unexpected call. And it turned out more pleasantly than if there had been a month's notice for preparation. His Majesty was properly impressed with the majesty of Skibo and was charmed by Louise's gracious and informal hospitality. But the hit of the afternoon was six-year-old Margaret, who had been gathering flowers along with her ever-faithful Nana on the lower terrace. Unannounced and quite unabashed by the royal presence, she marched up to the King and being the good American republican that she was, without benefit of a curtsey, held out one rose and said, "Here. This is for you." And then taking another from the bunch she held in her hand, she said, "Give this to the Queen when you get home." Edward was delighted. He picked her up, put her on his lap and kissed her soundly.[10]

If the Carnegies brought much that was American to Skibo —their easy informality of entertaining, their Fourth of July celebrations with the Stars and Stripes sewn to the British ensign—they were, nevertheless, sensitive to and made an effort to respect Scottish customs. They dutifully observed the strict Sabbatarianism of this Calvinist land. There were no parties given on a Sunday, no piper skirling the pipes on that morning and no card playing in the evening. Instead, the family, the domestic staff, and any guests who might be in residence would gather in the great hall after the evening meal and like any good Scottish Presbyterian family, sing favorite old hymns. Travel beyond the estate was taboo except in an emergency. In one letter, Carnegie wrote to Morley regarding the latter's travel plans to Skibo, "You may have trouble getting a Berth . . . Monday eve the tenth might be crowded—so why not Saturday eve & arrive Sunday noon. We can send for you even that day without shocking the people too much when we can urge a Cabinet official's needs. For an ordinary man this palliation would be insufficient. . . ."[11] Critic though he was of all established churches, Carnegie even made the supreme sacrifice of going to the cathedral in Dornoch on the first and last Sunday of each summer's residency—the minimum in church attendance that was expected of all Scottish lairds. And although cartoonists on both sides of the Atlantic invariably depicted Carnegie dressed in the kilt, complete with sporran and glengarry, there is only one recorded instance in which he was actually seen so garbed. He told Louise afterward that he felt damned uncomfortable the whole time. Carnegie was much too respectful of the old Scottish protocol which restricted the wearing of the kilt to those born north of the river Tay to assume the dress of a Highland laird, along with the title.

Louise Carnegie also had little difficulty in adjusting to Scottish mores with a few minor exceptions. She refused to drink milk in her tea and was delighed when King Edward at their impromptu tea also asked for lemon instead of cream. She also insisted that Scottish Sabbatarian rules must bend enough to allow her to have her usual late-morning swim and for Andrew to play a round of golf on their own estate if he so desired. Otherwise, living within a foreign culture presented no difficulties. From the moment she had come north of the border as a bride, she had felt almost as much at home in Scotland as in New York.

She continued to find it difficult, however, to accept the

steady flow of visitors that came to Skibo throughout the summer months. Outwardly gracious and relaxed, she seemed the model of the accomplished hostess, who managed a household staff of eighty-five servants with the efficiency and dispatch of a seasoned troop commander and who warmly greeted each arriving guest as if this were the one person in all the world that she most wanted to see at that moment. But inwardly, the old tension was still there, even if few others could detect it—least of all, Andrew. Old Uncle George Lauder had been sensitive enough, however, to appreciate the kind of demands that constant entertaining made upon her. Several years before, while the Carnegies were still at Cluny, he had written her a private letter in which he warned of the toll he realized she was paying:

> Now pay great attention to what I am going to say. Your dear husband Naig is drawing a large bill on futurity and I think you are a partner in that Bill, which no one will be able to pay but yourselves. So stop in time, you are both making work that would be sufficient for another strong couple, entertaining a relay of visitors the year through. And not ordinary men but all men of talent and ladies of accomplishment is no ordinary work. You cannot stand it, no man or woman can stand it. Excuse me for intruding my opinion in your domestic affairs and believe me it is because I wish you a long life of happiness in this world and a happy meeting in the next.[12]

During those first three summers at Skibo, even though great demands had been placed upon her to supervise the new additions, she had found herself much more relaxed than later when the renovation was complete and the crowds began to come again in even greater numbers than at Cluny. Her frequent letters to her mother had been a necessary escape valve for the inner pressure she felt as a bride at Cluny, but she no longer could unburden herself in that way. Only occasionally, in notes to her secretary, Archibald Barrow, or once in a while in a letter to her husband's solicitor, John Ross, who had become a close friend, would she give some indication of the weariness and nervous exhaustion she frequently felt.

It would be a mistake, however, to suggest that Louise Carnegie did not enjoy life at Skibo. There were the late-morning swims, the few hours in mid-afternoon alone in her room to rest, read, and write letters, and the late evenings and nights with Andrew all to herself. She even enjoyed much of the entertaining she was obliged to do. At the dinner table there would often be some of the most important people in Britain and America, the conversation was never dull and frequently fascinating as it ranged over politics, literature, and the latest projects for establishing world peace. There were even times when she seemed to be as exuberant and as vivacious in the company of others as Andrew always was—moments of what she called "high jinks"—when she herself would ask some neighbors in to play "puff billiards," or with a few friends and Andrew aboard their yacht, *Seabreeze*, she would suddenly burst into song, "Charlie, Charlie, wha wadna follow thee, King o'oor Highland hearts, bonnie Prince Charlie!"

Above all, there were the wonderful moments alone with Margaret, attended only by the devoted—perhaps too devoted—Nana, when they would walk together in the gardens and Louise would teach her daughter the names of the flowers, or when she would take Margaret in their open carriage into Bonar Bridge or Dornoch for some short errand.

Margaret was a lively, precocious child, growing up in a wonderfully strange, adult world, accepting the extraordinarily illustrious visitors to her home as her ordinary companions. She would remember sitting on Rudyard Kipling's lap while he told her fascinating "just so" stories of jungles and tigers and a mongoose, Rikki-Tikki-Tavi. She would never forget Kipling's eyes—the most fascinating eyes she would ever see—which held her as spellbound as did his stories. Of all of her father's guests, however, her favorite was Booker T. Washington. He was the only black man she knew, and there was a warmth in his Southern speech and in his manner of treating her as a real person, not as a pretty little doll, to which she openly responded. Perhaps they instinctively felt, each for his or her own reason, that they were allied aliens in a world that was both adult and white. But her father's favorite guest, John Morley, was not hers. Never having had a child of his own, Morley was as ill at ease with children as they were with him. Morley was no "Old Shoe" to Margaret. Quite the contrary, they both felt the relationship pinched and were quite content to keep a respectful distance from each other.[13]

Soon after the renovating of Skibo was completed, Carnegie was able to remedy the one defect, in his opinion, that Skibo had—its lack of a natural waterfall. He persuaded the Duke of Sutherland to sell a vast tract of moor land to the north and west of the castle which nearly doubled the size of the original estate, and by extending to the River Shin, provided Skibo not only with even better salmon fishing than was to be found in the Evelix but also gave the estate one of the finest waterfalls in northeastern Scotland. On this newly acquired land, high upon the moors there was a stone cottage, called Achindinagh, and here the Carnegies had occasionally stopped for a picnic while out inspecting their newest acquisition. Louise had loved and treasured every one of these brief moments.

In the spring of 1904, as the Carnegies prepared to leave for Scotland, Louise looked ahead with some apprehension to what promised to be their busiest social season yet. She finally revealed to Andrew her weariness and the strong desire that she, he and Margaret might this coming summer get away by themselves if only for a short time. Three weeks without visitors, without a corps of servants, and without constant entertaining would be "heaven on earth" for her. Only then did

Carnegie realize the kind of strain that Louise must be living under each summer. And as usual, when once informed, he sprang into action. Orders were cabled to Scotland to have Achindinagh refurbished and made livable. They would try out a three-week retreat this very summer to see how they liked it.

In the meantime, there were commitments already made that had to be changed. Carnegie wrote Principal James Donaldson of St. Andrews, "She has no long Holiday term as you lucky Scotch Professors and I have. Skibo I call one uninterrupted playtime. Not quite so for the 'Boss of the show' however—woman's work is never done."[14] And to Morley, he was more explicit. "The family go into retreat July 22—to Aug. 10th away up the high moors. . . . Cottage on the Shinn. Madam thinks higher air for two weeks best for Baba [Margaret's family nickname]. She also gets a rest preparatory to shooting season. It is an experiment we are to try. I think a wise one. She wished to postpone it until next year when you wrote you could come but I insisted you could & would come later or *sooner*. Perhaps you can spend some time here before 23d July your Edinburgh meeting."[15]

The experiment proved a great success, and thereafter a three-week retreat on the moors became a regular part of the Skibo summer. For Louise it was the best part of the summer; indeed, the best part of their lives. For three weeks they lived a simple life as an ordinary upper-class family, only she, Andrew, Baba, and of course, Stella, along with Nana and two servants. It was the kind of home life she had imagined as a romantic young girl, dreaming of her future as a wife and mother. Here Carnegie had ample time for his writing. He finished his biography of James Watt and began in earnest a project he had long had in mind but had made only desultory starts on before—the writing of his personal memoirs. In 1910, finding Achindinagh a bit too primitive and cramped for space, they built a larger lodge, named Aultnagar, in the more wooded and mountainous area of the estate, overlooking the Shin valley and more reminiscent of Cluny. But the more spacious quarters were solely for their own comfort—no visitors allowed.

Here in their retreat, Margaret and her father were able to become really acquainted. The weeks of lonely isolation on the moors in the company of her parents and aunt did not seem to Margaret as empty of companionship as they would have seemed to most children. She delighted her father with her precocity and quick wit. "I must tell you Margaret astonished me the other night by repeating the Seven Ages of Man," the proud papa wrote to Morley. "All perfect & in fine style. She is developing fast—puzzles her mother about certain things in Holy Writ now & then that gives Madam some anxiety. She does her best & I say, all right, Lou—I'll not give you away. Do the best you can—but remember she'll find out the truth before long for herself & Lou agrees. Yes, she won't rest until satisfied."[16]

The old radicalism of her Dunfermline forebears also manifested itself anew in Margaret. Carnegie reported to Richard Gilder that Margaret's "chief work is making up parties who never had a motor ride and taking them as her guests—all our servants in turn—picnics also provided. We encourage her in doing for others, already the young socialist crops out. . . . Why should some be rich and others poor, why do we invite rich people and give them everything when they have plenty at home—and the poor haven't. Questions easier to ask than answer. Her babble for a day would enrich even the Century columns."[17]

If Margaret delighted her father, she in turn was delighted with him and with the attention he could give her while at Achindinagh as he could nowhere else. Remembering the hours he had spent as a child in the Lauder grocery store in Dunfermline listening to the stories and the history his uncle told and how much those sessions had meant to him, Carnegie retold all of the Celtic legends and fairy stories, taught her long passages from Shakespeare, the poems of Robert Burns, and encouraged her to question everything that came to her attention. In spite of what must have been a difficult childhood for her, Margaret, because of those retreats, would cherish the memory of a special relationship with her father, which most children, with fathers much nearer their own age, would never know.

The retreat into the moors became especially precious to the Carnegies after Margaret became seriously ill at the age of eight. What appeared at first to be only a sprained ankle, suffered at Skibo in the late summer of 1905, did not heal properly. When the specialists in Edinburgh examined her leg, they suggested that Margaret might be afflicted with a serious bone infection. Upon the return of the family to New York, Margaret's entire leg was put in a plaster cast, and the orthopedists could give no assurances that a cure could be effected. Carnegie, who had so rarely been touched by the normal vicissitudes of life, was quite unprepared emotionally for any tragic anxiety. "We are having the first active pang of grief," he wrote Morley. "Baba's leg in plaster cast. . . . It may be only the result of a sprain, but I fear Edinburg [*sic*] specialist Prof. Stiles believed it more serious. We are so anxious. Can't tell for three or four months & naturally fear the worst. It is terrible."[18]

A later diagnosis was much more hopeful. It proved to be not the deterioration of bone that they feared but, as Carnegie wrote Morley a short time later, an affliction "of a gouty nature & not as the Edinburg [sic] specialist feared. This seems decided, a heavy weight is lifted from our hearts."[19] Even so, Margaret had to wear splints for nearly three years. With the child's illness, the always overly solicitous Nana truly took charge and with the best intentions in the world, nevertheless smothered Margaret with care. Many of the Carnegies' friends and relatives felt that the child was treated as an invalid far longer than her actual condition warranted. It was only in the retreat on the moors, when Louise could give her own full attention to her daughter that the little girl found some measure of freedom from Nana's constant care—one more reason for Margaret's happiness at Achindinagh.

Carnegie had other concerns in these years in addition to his anxiety over his daughter's health. He was finding the business of philanthropy in many respects far more difficult than the business of steel. At first, it had all seemed so easy. Decide on which areas would be of the greatest benefit to humanity, create a board of able and dedicated trustees to supervise the management of the trust, then write the check and sit back and revel in the plaudits of the multitudes. This was how he said the gospel of wealth would work, and this is how it did work—initially. In addition to the fifty millions he gave for community public libraries, he created two great institutes in the United States, three major philanthropic trusts in Britain and four in the United States. Three of the American trusts, including the Carnegie Hero Fund with separate endowments in ten other countries, were concerned with the promotion of world peace. He also had built three imposing structures, "temples of peace," as he liked to call them, one to house the International Court of Arbitration at the Hague, another the Pan American Union in Washington, D.C., and the third for the Central American Court of Justice in Cartago, Costa Rica.

The plaudits came, and how sweet it all was! "Never so busy, never so happy," Carnegie would frequently write Cousin Dod or Friend Morley, as he went around Britain collecting the Freedom of fifty-seven cities—an all-time record which not even Winston Churchill was later able to surpass. There was even the heady excitement of competition about this business of giving. As trade journals had once carried monthly production figures of the steel plants of the nation, pitting Carnegie Steel against Cambria or Illinois Steel, so now the daily press frequently ran stories of who was ahead in philanthropy, Carnegie or Rockefeller, and Carnegie always came in first. In 1904, *The Times* of London reported during the previous year Carnegie had given twenty-one million, Rockefeller only ten. By 1910 the New York *American*'s box score read for total lifetime giving: Carnegie $179,300,000; Rockefeller, $134,271,000. Whether or not these figures were accurate made no difference to the American public. The little Scotsman was far ahead.

But by that time, the whole business had turned sour for Carnegie. He had not anticipated the difficulties the establishment of each of these foundations would create: the angry outburst from the elite for providing for free tuition at all four of the Scottish universities, the petty squabbles over the architectural plans for the Peace Palace at the Hague, the uproar that resulted from Carnegie's excluding all sectarian colleges and universities from his Teachers' Pension Fund. Frequently, the boos would drown out the applause, and to Carnegie's amazement he found himself far more an object of criticism in giving away his wealth than he ever had been in acquiring it—even taking into account the tragic Homestead Strike. How a man made his fortune was his own business, but when he tried to give it back to society, it became everybody's business.

So life was not all sunshine at Skibo, and sometimes Carnegie felt that he was working harder now at philanthropy than he ever had in industry. He wrote to John Ross, "The final dispensation of one's wealth preparing for the final exit is I found a heavy task—all sad. . . . You have no idea the strain I have been under."[20] And once in giving a speech in Edinburgh, he departed from his prepared text to say bitterly, "Millionaires who laugh are rare, very rare, indeed."[21]

Most discouraging of all was the fact that as fast as he gave away his millions, more millions continued to accumulate on those remaining 5 percent gold bonds. By 1911, he had given away $180,000,000, but he still had almost that amount left. In his celebrated article on "Wealth" for the *North American Review* in 1889, Carnegie had enunciated his most-often-quoted aphorism, "The man who dies rich, dies disgraced." Now it apppeared, in spite of all of his efforts, he would die in disgrace after all. His good friend Elihu Root had tried to cheer Carnegie by writing, "You have had the best run for your money I have ever known." And so he had, but the capitalistic system at 5 percent could run as fast as he. Root had a simple solution. Why didn't Carnegie set up a trust, transfer the remainder of his fortune, after making generous allowance to his wife and daughter, to that trust for others to worry about and so die happy in a state of grace?

Carnegie did just that. In November 1911, he created the Carnegie Corporation of New York to which he transferred the bulk of his remaining fortune, $125,000,000. He could now say a final Amen to the Gospel of Wealth, turn his full attention to directing recalcitrant Presidents and ex-Presidents of the United States, British kings and prime ministers, German emperors and chancellors toward world peace. He could also once again thoroughly enjoy the sunshine of Skibo.

But the sunshine proved of short duration. The Carnegies were still at their retreat at Aultnager on 1 August 1914 when Germany declared war on Russia followed immediately by France's declaration of war on Germany. Three days later, German troops invaded Belgium on the road to Paris, and Britain promptly declared war to support her ally, France, and to protect the neutrality of Belgium. For Carnegie it was as if the very planet had cracked–suddenly, senselessly, without warning.

On the very day he received word that the Great War had begun, Carnegie had just finished his autobiography–the self-portrait of a self-satisfied man. The last page of his memoirs was entitled "The Kaiser and World Peace," and it concluded with Carnegie's account of his meeting the German emperor the year before and presenting him with "The American address of congratulations" upon the silver anniversary of his peaceful reign.

> As I approached to hand to him the casket containing the address, he recognized me and with outstretched arms, exclaimed: "Carnegie, twenty-five years of peace, and we hope for many more."
> I could not help responding: "And in this noblest of all missions you are our chief ally."

It was a happy ending to a happy book. Like *Triumphant Democracy*, his *Autobiography* had been "all sunshine, sunshine, sunshine," But now the sun was suddenly blotted out. Carnegie had to add a postscript to the completed manuscript:

> As I read this today, what a change! The world convulsed by war as never before! Men slaying each other like wild beasts! I dare not relinquish all hope. I see another ruler coming forward upon the world stage, who may prove himself the immortal one. . . . Nothing is impossible to genius! Watch President Wilson! He has Scotch blood in his veins.[22]

Even in this dark moment, Carnegie could not really believe that all was lost. But how hollow his old "onward and upward" cry sounded now as Europe plunged downward into Hell. John C. Van Dyke, who edited Carnegie's memoirs in 1920, added his own postscript to the book: "[Here the manuscript ends abruptly.]" And so, in reality, did Carnegie's life.

The Carnegies hurried back to Skibo on the day that Britain declared war. These last few weeks were the only truly dark days Carnegie had ever known there, even though nature, once again out of tune with man, mocked them by continuing to give them one bright, sun-filled day after another. They tried to continue the familiar old routine. Morley, who, true to his pacifist principles, had resigned from the Cabinet, came for his usual late-summer visit, as did a steady stream of other

English and American visitors, including young Nan Carnegie, the granddaughter of Andrew's long-dead brother Tom Carnegie.

But Skibo was no longer Schytherbolle, the fairyland of peace. With each passing day, the horrors of the war were brought closer to his unreal Camelot: the growing casualty lists printed daily in *The Times* and *The Scotsman;* the tears that were shed as young men on the staff at Skibo and from the tenant farms left for that bloody front in France. Carnegie wrote to John Ross in mid-August: "Our horses, traps &c commandeered—our territorials, ditto. All, the household servants included, steadily at work, sewing & knitting for the Army. It is all too sad to contemplate but we can indulge the hope that out of this eruption there is to spring the resolve to form an organization among the nations to *prevent war* hereafter. . . ."[23] It was now the only indulgence that Skibo could offer.

The Carnegies made their annual farewell to the household staff on the morning of 14 September, about a month and a half earlier than was their usual custom. As might be expected, the parting was more tearfully emotional in this sad autumn, but no one, least of all Carnegie, realized that this was his final farewell to beloved Skibo. Driving by auto to the train station at Bonar Bridge, the Laird of Skibo had an opportunity for a good view of his estate. The trees that he had planted along the roads were in brightest autumnal foliage, and the hills beyond were at their deepest heather purple, the traditional color for royalty—and also for mourning.

The last night in Britain was spent at a hotel in Liverpool. Morley, as he had frequently done in the past, came up to be with the Carnegies on the eve of their departure. The two old friends sat up far later into the night than was their custom, talking over many things. Morley was in a state of deep depression; Carnegie, as usual, was resolutely hopeful, outlining his plans for a world peace council which he intended to persuade President Wilson to call, as soon as he got back to America.

In the crowded Liverpool dock terminal the next morning the farewells could only be hurried and brief and so the parting was made easier for Carnegie and Morley. Nor could there be any lingering backward look as the ship sailed down the Mersey and out to sea, for fog lay heavy over the entire Irish Sea, and Britain was quickly blotted from view. The days of sunshine at Skibo were over.

Carnegie's closest friends and associates, when the first news of the outbreak of the war had come, had been fearful that Carnegie might not survive the blow. Yet after the initial shock was over and the Carnegies were aboard ship and headed back to America, Carnegie, with vigor, energy, and even optimism, had begun to lay plans for the American government to initiate a peace conference and then to assume leadership in achieving an effective international organization of peace. In October, he wrote an article for *Independent*, entitled "A League of Peace–Not 'Preparation for War,'" which proved to be his last published article. "One thing is certain," he concluded, "peace upon earth can never come from 'preparation for war,' hence let us discard that fallacy and try other means. It is submitted that a League of Peace embracing the chief nations is worthy of consideration."[24]

Carnegie was delighted with Wilson's statement, "There is such a thing as a man being too proud to fight," and even more pleased when the President sent his close friend, Colonel E. M. House, to Europe to visit with the heads of state of each of the warring powers as preparation for what Carnegie wanted–a general cease-fire and a world conference to settle all differences. How wrong, Carnegie realized now, he had been to support Taft in 1912. The American people in their infinite wisdom had elected the greatest President since Lincoln, and he in turn, in his infinite wisdom, had selected the greatest Secretary of State America had ever had, with the possible exception of Elihu Root. And to think Carnegie had once seen William Jennings Bryan as a wild demagogue who would destroy the Republic. Carnegie's trips to Washington were at least as frequent as they had been during the presidencies of Roosevelt and Taft, but now he got a warmer reception than he had ever received from a Republican administration.

But slowly hope seeped away. House's grand tour of the war capitals of Europe was a fiasco. Wilson grew more belligerent, and Bryan, as Morley had done in Britain, felt obliged to resign from the President's Cabinet. The dogs of war, once unleashed, could not be chained by hopeful talk, and the drums were even beginning to beat in peacefully isolated America. Carnegie started to age very fast, and by the time the United States finally entered the war in 1917, he was a sick, tired, old man of eighty-two, finding it difficult to write even a letter to Morley. There was nothing to do now but sit and wait—wait for this madness to run its course; wait for the killing to stop; wait for Wilson, as victor, to impose his League of Nations upon a devastated Europe, and for Carnegie personally, wait until he could get back once again to Skibo. Only then might there be a resurrection of hope and a renewal of life.

Louise Carnegie did her best to sustain her husband in these last dark years. There were the winter cruises on their yacht in the Florida coastal waters, which she secretly detested, and in 1916, after two unsatisfactory summers, first in Maine and then in Noroton, Connecticut, she at last found the best possible substitute for Skibo, a beautiful, very large stone mansion in the Berkshires, near Lenox, Massachusetts. Built by Anson Phelps Dodge in 1892, Shadowbrook was no simple, little summer cottage. Only George Vanderbilt's chateau, Biltmore, in Asheville, North Carolina, exceeded it in being America's largest private residence. With its sixty rooms, it was far larger than the Carnegies needed, now that there were few visitors, but she knew Andrew would not be content with a modest summer home, and the setting, overlooking Lake Mahkeenac, was beautiful. Carnegie gave his approval to her find and said it would do quite nicely "until we return to Skibo."

Her husband's slow fading out of life occupied much of Louise's thoughts and energies, but not exclusively. There was also their daughter's rapid emergence into life to consider. Margaret, who had graduated from Miss Spence's School in 1916, was now twenty-one, a beautiful and gracious young lady. In the summer of 1918 she had fallen in love with a young ensign, Roswell Miller, the brother of a close friend of hers at Miss Spence's School, and the son of the former president of the Chicago, Milwaukee & St. Paul Railroad, whom Carnegie had known many years before. Compared with her parents' extended engagement, Margaret's romance was a whirlwind affair. Upon the Carnegies' return to New York from Shadowbrook in early November 1918, they were informed by Margaret and Roswell that they wished to marry. The old man wept but gave his consent. It was agreed that the engagement would be announced at Andrew's birthday celebration on the 25th of November.

Carnegie's eighty-third birthday came two weeks after the Armstice was signed and the guns were at last silent. It was the happiest birthday he had known in years. The betrothal announcement was made to the assembled company of friends. Margaret and Roswell would be married the following April 22nd, Andrew's and Louise's thirty-second wedding anniversary. Then finally the enlarged Carnegie family could all return to Skibo.

The wedding day, as Louise wrote in her diary that evening, was a "Glorious bright spring day." Frail and feeble as he was, Andrew insisted upon escorting Margaret down the grand staircase of their home at No. 2 East 91st Street to the altar that had been erected in the conservatory. It was Carnegie's last public appearance. "Margaret made a very lovely bride," Louise continued. "Andrew so well and alert. He and I gave Baba away and later we walked down the aisle together. After greeting the bride and groom he went upstairs and rested."[25]

Margaret and Roswell left after the wedding luncheon in his new Stutz enclosed car, with the indomitable Nana, who insisted, to no one's great surprise, upon accompanying them to look after her little girl's clothes, seated stiffly in the back seat along with the luggage. And now Louise was quite alone with Andrew. They played backgammon that evening.

Inevitably, the question of the return to Skibo came up. Now surely that the war was over, they could go back. But Louise had to tell Andrew that it would not be possible this first spring of peace. Transportation was too uncertain—and dangerous with the waters around Britain still mined. And there were serious food shortages. Better keep to Shadowbrook for this summer. But she found it difficult to face the hurt look in his eyes, so she tried to keep the pretense of their return alive for just a little longer. "But surely next year, when things are back to normal we can go." Carnegie looked up at her with the same sharp, penetrating glance that used to strike terror in the hearts of business associates who were trying to fool him. "There won't be a next year for me," he said abruptly. Neither of them ever mentioned the return to Skibo again.[26]

The Carnegies went back to Shadowbrook in May. Margaret and Roswell motored up from his family home in Connecticut, full of excitement about the house they had found in Princeton, where Roswell in the fall would resume his studies which had been interrupted by his service in the navy during the last year of the war. Then the long days of summer dragged slowly by. There was no fishing for Andrew this summer, only quiet, patient waiting. The only comfort for Louise was that Morrison, Carnegie's devoted valet, had been released from military service and was there at Shadowbrook to care for Andrew.

On 9 August Carnegie was stricken with pneumonia. Both Louise and he knew that the end was near, but Carnegie was so very tired that he welcomed it. A little after seven on the morning of 11 August, he quietly died.

Of all of the hundreds of letters and telegrams that came to Shadowbrook from around the world, the one that touched Louise Carnegie the most came from John Morley. "How little when we last said goodbye at the Liverpool Station, could we suppose that we were to meet no more, and that the humane hopes we had lived in, and lived by, were on the very eve of ruin. . . . My days of survival cannot be very prolonged, but they will be much the more dull now that the beacon across the Atlantic has gone out."[27]

There were many besides Morley who would find the world a much duller place without Carnegie. Most of all, the people at Skibo felt the bright beacon had gone out. His saddened tenants cut a great slab of granite from a quarry on the estate, had it sent to Glasgow to be carved by a sculptor into a Celtic cross, and then had it transported across the Atlantic to mark his grave. And on a hill above the road to Bonar Bridge, there was erected a simple stone cairn, without plaque or name on it. But all good Celts in the region knew, as their ancient Pictian ancestors would have known, that this was to honor a most worthy laird, who had brought to his land and people great glory and renown.[28] No words were needed.

V

Skibo, the Gracious Dowager: The Louise Whitfield Carnegie Years

1920 – 1946

Louise Carnegie's thoughts turned often to Skibo during the autumn of 1919, the first months of her widowhood. It did not take a lovely wreath of deep lavender heather which she received for her husband's grave from the tenants of Skibo in September to remind her of those purple moors, so distant in both place and time. It had been five years almost to the day since she, Andrew and Margaret had driven for the last time together down that avenue of trees and out the main gate which the staff—and that time for them and not for some distinguished guest—had decorated with the old familiar sign, "Will ye no come back again."

Louise did want to go back again, and would go back next summer for sure, painful as she knew the first return without Andrew would be. But whether she could keep Skibo, could continue the annual migrations remained to be seen. She knew that this is what Andrew would want her to do and what she herself would very willingly do, but so much depended upon Margaret and Roswell. Her son-in-law had never seen Skibo. It had, at least as yet, no meaning to his and Margaret's life together. Would they want to build Skibo into as an essential part of their lives as she and Andrew had done? If not, then she would have to part with Skibo, for she had no wish to continue the migrations alone.

But nothing could be decided at this moment. They would all go back to Skibo next summer—Roswell and Margaret had agreed to that, and then a decision could be made. There was one thing of which Louise was certain, however, and that could be settled now. She had no desire to return ever again to Shadowbrook even if she did not keep Skibo. The Berkshire home had its own great beauty; it had served the Carnegies well as the best possible American substitute for Skibo; but Andrew's and her last home together had only sad memories for her. Shadowbrook must be sold and as quickly as possible. The estate was put on the market in the spring of 1920, but there was no immediate offer forthcoming. It proved no easy task to dispose of a mansion of that size with ever-rising property taxes. It was not until October 1922 that Shadowbrook was finally sold to the Jesuits of the Province of New England for a fraction of its cost to the Carnegies. The Jesuits intended to convert it into a seminary for 150 students, and it was ideally suited for that. Louise must have wondered if Andrew, with his views on churches and particularly missionary organizations, would have found ironic humor in that. From Skibo, which long ago had been the Roman Catholic archbishop's palace, to Shadowbrook, now to be a Roman Catholic seminary—the Church giveth and the Church taketh back. Here was a neat historic circle that might have amused him.

When the story broke in the press that Mrs. Carnegie had just put Shadowbrook up for sale, a rumor quickly circulated that she was also looking for a buyer for her home on East 91st Street and that she and the Millers would then move permanently to Europe, dividing their time between Scotland and southern France. Louise's private secretary, Archibald Barrow, in an interview firmly squelched such stories. Mrs. Carnegie and her daughter and son-in-law had no intention of moving permanently out of the country. Quite the contrary.

Not only did Louise Carnegie have no intention of selling her home in New York, she had but recently purchased the former residence of George L. McAlpin at 9 East 90th Street, directly behind her home on 91st Street. Roswell would be graduating from Princeton after one more year of study, and it was Louise's great hope that she could persuade the Millers to accept the McAlpin house as a gift and make their home there.

How she would have welcomed them into her great mansion, but she knew that would not and should not be. She wrote to sister Stella from Princeton where she had gone to be with the Millers soon after Andrew's funeral, "My life for the past six years has been one of continual change, so I accept things as they come–sustained by the life within and learning what it means to stand alone. . . . Margaret and Roswell are dear; no two children could be dearer and more thoughtful, but I am inevitably an outsider. This is the lesson we all have to learn and we do not truly live until we have learned it."[1] Poor Stella, who had never really had a life of her own, having lived with her mother until the latter died, and then with Andrew and Louise, was a curious sounding board for Louise to have chosen in expectation of a sympathetic vibration. It was, nevertheless, a wise and necessary note to sound, even if Louise could not be sure that the Millers would appreciate the fact that she really did mean them to have their own lives, independent of hers.

But at least they would all be together this summer at Skibo even if there would not be the customary late May departure from New York that Louise had originally planned on. Margaret and Roswell were expecting their first child in June and assuming that the baby should be healthy enough to undertake a trans-Atlantic voyage so soon after birth, it would be at least late July before they could depart. But what was a short wait of only two months after the six long years of waiting she had already endured. And how wonderful to have to postpone Skibo this time in order to await a new life to begin as compared to the long, sad delay while an old life slowly ended.

The extra time would give Louise an opportunity to see her husband's autobiography through the final stages of preparation for publication. John C. Van Dyke, professor of history at Rutgers University and long a close friend of both her and Andrew, had agreed to edit the manuscript, and it would be published by Houghton Mifflin Company. Louise herself wrote the preface to the book—her public tribute to Andrew Carnegie.

In early June the Millers came up from Princeton to stay with Louise during the final days of Margaret's pregnancy. In the very early hours of the morning of 17 June 1920, the baby was born, a daughter whom the Millers named Louise Carnegie Miller. Nearly two decades later, Louise Carnegie wrote a birthday letter to her granddaughter describing her memories of that happy day:

> I see the Nurse coming to me at dinner time the evening before saying I must not think of staying at home but must go to the Wednesday evening service as usual—"it would be better for Mrs. Miller."
>
> The coming home at 9:30 to see Daddy leading Mummie downstairs from the third floor to the Family Library on the second floor, Mummie clad in a pale blue tea gown like a Madonna. Then Daddy playing Scotch songs on the gramophone. . . . Then Dr. Thomas leading Mummie into her own room, while Daddy and I retired to my sitting room and rested on the two sofas until the wee sma' hours—when Nana came to the door and said, "I hear the baby crying!" and the wave of joy and thankfulness that went up from my heart that now this new life had come into our home! And what a blessing that new life has been to us all, all our days, ever since![2]

Baby Lou, who was soon to be given the nickname of Dede, proved to be a very healthy child indeed. Margaret had been only nine weeks old when the Carnegies had taken her across the Atlantic, but her daughter outdid her by three weeks in becoming a seasoned ocean voyager. Louise Carnegie left New York first in order to get Skibo opened up and in full operation again prior to the arrival of the Millers, who with the baby and the ever-faithful Nana, delighted to have a new charge, would arrive a few days later.

Louise Carnegie with sister Stella arrived at the Bonar Bridge station on the afternoon of 30 July 1920. There was the automobile waiting for them as if it had been keeping vigil for six years. And then the familiar ride along the firth, through the gates, and up to the castle entrance. Mr. Hardie, the factor, and his wife were at the door, and behind them, Mrs. Mark, the Carnegies' housekeeper since their first summer at Cluny, thirty-two long years ago. Louise was home again in Skibo. After a quick inspection of the castle and the gardens, Louise sat down and wrote a long letter to Margaret in the hope that it would be received before the Millers left New York.

> Darling,
>
> I'm at Skibo! just think of it! and everything in the house is exactly as we left it six years ago. We might really have been gone only six days from the way everything looks, except that in the old nursery there is a little pink bassinette waiting for its wee occupant!
>
> I have had a hasty run through the garden—the strawberries are not over yet—the cherries still hanging on the east wall big dark red ones, not many but still a few and oh! so sweet and the gooseberries and raspberries are just about ripe—they are all so wonderful! . . . Well, you will see it all for yourself soon. Will it be Monday or Tuesday I wonder. I fear the latter. Mr. Hardie says there is lots of grouse and the dogs are being trained, and

> there are lots and lots of rabbits. My! won't they taste good—rabbit pie for instance—so there will be lots of sport for Roswell. . . . We have had sunshine and showers all day just like April—but the sky is wonderful. The house is warm and comfortable and of course open fires. . . .
>
> Mr. Hardie gave me your dear letter on arrival. Oh! darling it was so sweet of you to think of writing.[3]

Louise's letter was designed to make Margaret as eager to get back to Skibo as she herself had been, for her daughter's letter which she had received upon her arrival had evidently intimated that Margaret felt a certain reluctance in being engulfed by her childhood again. Louise felt it necessary to add to her rhapsodizing over the delight of coming home to Skibo a more serious note regarding their future relations as a family:

> But just here I want to set you right about something you said in your letter. I don't expect you to have my viewpoint—and I wouldn't have you feel one day older than you are. I want you gay and young and frivolous and bent on having a good time both of you together and it makes me happy to help you have a happy young time. I say this because I know you have a heart under it all. You are bringing *your* family to *your* home and I don't expect things to go on in the old way and don't want them to. You are the daughter of the house and and I am only happy when you are happy. We both have our separate lives to live but it is a comfort to live them near each other. This is not a sermon!!! Now hurry up and come home and don't take cold any of you—it is a terrific change from N.Y. . . .[4]

It was a loving and endearing note, but also self-contradictory with its message of assuring Margaret that her mother rationally understood her daughter's desire for independence, and yet at the same time emotionally giving voice to the all-protective mother, urging her children to button up their coats and hurry home to her.

If Margaret had certain misgivings of what lay ahead, it is hardly surprising. Skibo did bring back to her wonderful memories of its beauty and charm, of a kindly and adoring father, old enough to be her grandfather, who told her grand stories about fairies and of brave Wallace and the noble Bruce. But Skibo had also meant growing up in a very adult world, of sitting quietly while elderly and learned men drearily discussed preferential tariffs and Herbert Spencer's gloomy philosophy and President Taft's futile arbitration treaties. Margaret had never known the wonderful egalitarian world of childhood. To the few children who came to Skibo to play, she was the daughter of the laird, who also happened to be, their parents told them in awe, the richest man in the world. And so she had stood apart, the little lady bountiful, distributing gifts to those much less fortunate than she, but never sharing in their giggling, whispered secrets, never being jostled in their rough and tumble games.

Her marriage had changed all that. Roswell all at once had given her a childhood and youth she had never before known. How wonderful that first year of marriage in that small house in Princeton had been—buying their wood and coal in midsummer as if it truly mattered that they could get a better price at that season. And Miller was such a wonderfully ordinary name. Being introduced to new people as Mrs. Miller did not immediately evoke images of public libraries and symphony halls and institutions that the name Carnegie always did. How much she must have enjoyed meeting Roswell's friends and fellow students as an equal and not have to lean down from a pedestal to shake their hands. Although they would have only one more year at Princeton, Roswell and she must not lose what she had found there. Much as she loved Skibo—and was sure that Roswell would too—she must now have felt a bit as if the old world were closing in on her again.

And Roswell did love Skibo—with extravagant enthusiasm. Like his mother-in-law thirty-three years earlier, he became a Scot as soon as he had crossed the border. Skibo exceeded even Louise Carnegie's advance notices of it. It was indeed what Andrew Carnegie had said, "a heaven on earth." Roswell even liked its formality, and to Louise's delight, insisted that the old rituals be resumed—the piper in the morning, the organ music at meal time, and dressing for dinner. Roswell had none of his father's-in-law respect for ancient Scottish taboos on dress. He delighted in wearing the kilt on all appropriate occasions and what a dashingly romantic and handsome figure he was in the ancient dress tartan. He quickly won the hearts of all the tenants of Skibo. Here was a laird's son of whom they could be proud.

In one respect, Roswell appreciated Skibo even more fully than Andrew Carnegie ever had, for he was an avid field as well as stream sportsman. Nothing pleased him more than being on the moors or in the forests, gun in hand, on hunt for the grouse or red deer. In the old days, Carnegie made it a part of the formalities of the season's opening day to stand before the assembled hunters and give them a little sermon on the evil of shooting to kill. This never stopped the hunt; Carnegie's guests took it in good humor, but it was not the kind of "Tally ho" send-off that any gamekeeper would appreciate. Now all that would be changed. This was one family ritual that would not be continued, and the Skibo gamekeeper would at last assume the importance he had on most Scottish estates.

For Louise, this first short summer back in Scotland was a decided success. Roswell quite clearly was immensely happy with Skibo, and the baby, with her ruddy, fat little cheeks, already looked like a native, wee bairn. But Margaret was not yet ready to give a commitment to the future, and she did not seem to be too unhappy that she and Roswell had to leave Skibo early. She was to be matron of honor at the wedding of Roswell's sister, Dorothy, who had been Margaret's classmate at Miss Spence's School and had introduced Margaret to her brother. Moreover, Margaret was eager to get the baby settled in their home in Princeton which little Lou had never seen. The Millers sailed for the United States aboard the *Aquitania* on 11 September with the question of the future of Skibo still unresolved.

Louise Carnegie, now that she was back at Skibo and had had Margaret and Roswell and little Lou with her for those glorious five weeks, knew more than ever that she wanted to keep Skibo. But could she continue to manage such a large estate and for how long would she be able to make the annual trip over? She wanted Skibo only if the Millers were a part of it too. She was not sure she could endure it alone. Was it fair to expect, even to want Margaret and Roswell to make the pilgrimage each year, to force them to build Skibo into their regular life together? And so Louise, who had always found indecision intolerable, vacillated between keeping and letting Skibo go during the first weeks after the Millers had left. Her almost daily letters to Margaret during this period reveal her indecision and how much she was torn between what she wanted and what she thought might be the wiser course to follow. During the first week she was at Skibo alone with Stella, she wrote:

> These coming weeks will be the real test whether I can keep this place. I find it is *you* darling and not the place that counts—and I cannot talk to you over the telephone from here! We shall see—I don't know if I can stand many more partings like this but the first is always the worst, and it will all work out well in the end. . . .
>
> I'm thinking of last Sunday morning, and the beauti-

> ful walk in the afternoon on the little moor. I am afraid I am making it harder for you, recalling all these times. . . . I was looking at the motto over the fireplace in the hall this A.M.—"hame is hame, but a hieland hame is mair than hame"—I don't believe you remember it, but isn't it true? Well, it doesn't do to get too much attached to *places*—we must move on and we all have work to do in the world.[5]

But then she would write the following day about plans for next year:

> I am busy in the house some days going over the linen closet, etc. Such beautiful new tablecloths—French—marked with your initials I find here, but I am not bringing them over just yet. You have two of your cribs here and I am getting the smaller one put up all ready for Lou next year!! . . .
>
> I have had a long sleep this afternoon after which I was busy framing pictures of you as a bride and with Baby and one of our family group. I want to leave them here. . . . It looks as if we intended to come back, doesn't it? Instead of packing photos to take home I am framing to leave here—heigh ho! I hope you will tell me just how you now feel in looking back over the summer. Does it seem to have been a big undertaking? Was it worth while? Would you dread doing it again? To know just how you feel about it now will help me very much—don't be afraid of my feelings—tell me honestly—weighing all the advantages and disadvantages how *you* feel about it.[6]

It was, of course, asking too much of Margaret to expect her to give the answer to the future of Skibo. Roswell's and her plans were too uncertain as to what they would do after he graduated and they left Princeton. She could not at this moment give any definite promises to her mother. And so Louise had to make the final decision herself, and in actuality, there probably had never been any real possibility that she would voluntarily dispose of what had been such an essential part of her life for nearly a quarter of a century. Skibo's old magic to bewitch and to hold its possessor still worked. By the end of September, Louise had made up her mind. She would keep Skibo no matter what the future summers might bring. She asked John Ross, their old friend and solicitor, to come up from Dunfermline and together they discussed what she might sell if she did not feel able to maintain the entire estate. And then she wrote Margaret of her plans:

> We are to give up all farming and reduce expenses as much as we can. If we can let Aultnagar with the Auchinduich shooting we shall do so. I want to sell all to the west of the Spinningdale road—this will let us keep Loch Bhuie and Lagain and the *Fairy Glen!*—but there would be no deer shooting nor salmon. Of course the Shin is the best part of our revenue but I don't want the responsibility and I do hope we may find a purchaser. . . . I am making my plans now to come back here early next summer, if you and Roswell find that your plans work out so that you can come over after R. gradutes—it will make me very, very happy. . . .
>
> If you and Roswell want to leave Baby with me and take a run on the Continent I shall be glad of this—but even if you feel you cannot come over and that you must take a house near New York—I could not be with you and my duties plainly lie here—if you cannot come I shall not stay so long. . . . I feel like a new creature now that I have begun some definite work and the awful thought of parting with Skibo has gone. I believe it was almost killing me. . . . The fast ships now would take me to you quickly if you needed me, and really you do not actually need me in the one sense, now that you have your husband and child and have your good doctor and trained nurse to call upon if necessary, and always your Nana. . . . and then you would always

> have Skibo to come to. It seems the right solution, doesn't it darling? . . .

The next paragraph of this letter indicates how anxiously the people of the Skibo estate had been awaiting Louise's decision. She told of the visit of her pastor in Bonar Bridge:

> Mr. Ritchie was here yesterday for luncheon and tea. I soon guessed what was on his mind—he wanted to know the result of Dr. Ross's visit. When I told him I was arranging to come back next year he gave a sigh of relief and spoke several times of how pleased the people round about would be. This is all very gratifying and I believe even if I have to curtail and live in a smaller way, which I shall earnestly try to do, that I can still bring a good deal of happiness into the lives of many people.[7]

With the issue settled, the remaining weeks of her stay at Skibo were a pure delight. "Auntie [Stella] and I are having lovely times together and she is very dear and thoughtful and I try not to be domineering."[8] There were the very amateur golf games together, and the drives on the moor and the picnics at Aultnagar, and virtually no visitors. Louise felt once again quite at peace with herself and her world. She returned to New York in late October refreshed and renewed. The decision had been the right one.

It was apparent when Louise Carnegie arrived back in New York that Margaret and particularly Roswell also felt that the decision had been the right one. They were already planning on coming to Skibo the next summer right after Roswell's graduation. So everything was working out just as Louise had wanted but hardly had dared to hope for. And the plans that Louise had made with John Ross to sell that vast extent of the estate extending west of Spinningdale to the Shin river were quickly abandoned. Louise knew that Roswell would forever regret the loss of the best salmon fishing and grouse hunting on the estate if that area were given up. Skibo would remain intact. Louise felt, now that she had the children at her side, she was capable of managing anything.

The return to Skibo in the summer of 1921 was postponed until late June. Since the Millers were also coming, Louise delayed her own departure until after the graduation. Margaret had also agreed to her mother's taking baby Lou up to Skibo as soon as they arrived in Britain while she and Roswell took a long and leisurely motor trip through southern England and Wales. It would be Roswell's graduation present and a second honeymoon for them both. Louise was delighted to be in charge of her little pride and joy. Upon arriving in Skibo on 5 July, she made the first proud entry in the Skibo guest book that had been made since that tragic summer of 1914 seven years before: "Louise W. Carnegie, Louise Carnegie Miller—arrived from New York & Princeton via Liverpool."[9]

It was a happy two and a half weeks that the proud grandmother had being a mother again prior to the arrival of the Millers on 22 July. Louise wrote to Margaret of one afternoon with the baby:

> Your darling daughter graciously consented to spend the afternoon with her grandmother and I have gone back 23 years and had a real old fashioned Sunday afternoon alone with her in the gardens. . . . I took Baby and sent Nurse off for a walk. We had a fine time. I gave her old roses to pull to pieces while I cut off the old calyxes, and this amused her for a long time, then she had a fine time throwing things out of her pram and Grandmother had a lot of exercise picking them up—it was as good as a game of golf for me physically! So much depends upon the way we look at things! Finally I spread a rubber rug under the big lime tree on the lawn and she sang Grandmother to sleep and tried to eat all the sticks she could find. It was half past five before I knew it and then I carried her in to bed in the best of spirits.[10]

Now that the Carnegie/Miller family was really established at Skibo, the visitors began to come again, but not in the great numbers of Andrew Carnegie's day. Old familiar faces reappeared: Margaret Blaine Damrosch and her daughters; the Lauder cousins from Dunfermline; George Adam Smith, the Principal of Aberdeen University, and his family; the Duke and Duchess of Sutherland; Grace Vanderbilt from New York, and one or two of the Old Shoes—Herbert Gladstone and his wife, and the Rudyard Kiplings. It was like old times, but the old times in moderation—just what Louise preferred.

The grouse hunting was exceptionally good that year, and when she saw the exultant Roswell returning from the shoot with his limit of the plump birds, Louise was more than ever happy that she had not sold off the western moorlands. Roswell had fallen an easy prey to Skibo's magic allurement. He was now a confirmed son of Skibo. No question now that he would be as eager to return every possible summer as was she. And that meant Margaret would be happy to do the same. Moreover, the Millers would take McAlpin house upon their return in the fall. Roswell had a position with an engineering firm in New York, so they would all be together after all—together but separate, as she firmly assured her children. It had become the best of all possible worlds for Louise Carnegie.

The family did not return to Skibo in the summer of 1922. Margaret was expecting the birth of a second child sometime in the early winter and did not wish to be away from her New York doctor during the early months of her pregnancy. For such a good reason as this, Louise was also quite content to stay in the United States. During June and July, she rented a summer home at West Hyannisport on Cape Cod. Compared with either Skibo or Shadowbrook, it was a very modest place indeed. The domestic staff that accompanied her were shocked by the cottage's smallness and simplicity, but it reminded Louise of that first summer retreat at Achindinagh, and she loved it. "No doubt there is plenty of grumbling going on behind the scenes, but luckily, I don't hear it," she wrote Margaret. "I am having the best holiday I have ever had in my life. I love every feature of my simple life and am able to forget all my worries in the freedom I now enjoy. . . ."[11]

This was, however, but one unusual, if pleasant, break in the regular summer pattern, and even there her thoughts must have been on next summer back at Skibo when the Millers would be coming with two children. Roswell Miller III, who was promptly and permanently nicknamed Robin, was born on 14 December 1922—a wonderful Christmas gift to them all.

Louise and Stella departed for Europe in April 1923. Louise had been invited by the French and Belgian governments to view the work of restoration that was going on in those two war-ravaged countries. Andrew Carnegie's Endowment for International Peace, for all of its brave hopes and its millions, had not been able to prevent World War I, but at least it could now spend some of those funds to repair the physical damages that war had inflicted. "The devastated regions saddened us very much—such havoc is beyond description—whole villages wiped out," she wrote home. "Rheims Cathedral is nearly roofed over, but oh! the destruction of the beautiful window. . . . We lunched at Soissons that day and in the P.M. went to Farnieres—where our Peace Foundation is rebuilding the school and a good portion of the village. The center is to be called La Place Carnegie. . . . Daddy's memory is revered here, and I want them to know that Daddy's family take an interest and wish to help them carry on."[12]

In Brussels, Louise was invited by King Albert to attend a formal state dinner that he was giving for the King and Queen of Spain. Albert reminded her of the dinner he had given for Andrew Carnegie ten years before when the Belgian Carnegie Hero Fund had been established. Now there were so many dead Belgian heroes, but these young men were sacrifices to the kind of heroism that Carnegie had hoped to make obsolete.

Back in Britain in late May, Louise spent some time in London visiting with old friends before heading north to Skibo. "John Morley came and lunched with us Saturday, the day after we arrived," she wrote from her hotel in London. "I see a great change in him; he is really very feeble and is very pathetic, but he is the same, dear friend."[13] It would be the last time they would meet.

The Millers, with Dede now two and Robin six months old, joined Louise at Skibo in early July. Now that the family was reunited at Skibo, a summer pattern could be firmly established that would prevail for the next sixteen years. Louise had promised Margaret that life at Skibo would not "go on in the old way," and it didn't. Although there was a pattern to their annual summer at Skibo, it was neither as formal nor as regular as it had been in the prewar days. The Fourth of July Fête Day was abandoned as was the Principals' Week. Louise still had some of the university principals up for a visit, particularly the George Adam Smiths of Aberdeen, but they came individually and not as a collective body. Nor did she any longer invite the entire Dunfermline Board of Trustees up for an annual meeting although William Robertson, the chairman of the board, usually came each summer. With the death of John Morley in September of 1923, there was no longer that group of friends especially designated as Old Shoes to make their frequent and lengthy visits. Old friends did make their occasional appearance–Aggie King, the Damrosches, Abram S. Hewitt, and of course the cousin Lauders, both the Dunfermline and American branches. But there were also the new friends of Roswell and Margaret and various members of the Miller family who now found their way to Skibo. The great notables of British literature and politics whom Andrew used to collect at Skibo as avidly as other men of wealth collected Renaissance paintings were now either dead or too infirm to make the trip. Other persons of distinction, to be sure, the Archbishop of Canterbury and his wife; Frederick Keppel, president of the Carnegie Corporation of New York; Helen Keller and her companion, Polly Thomson; and Burton Hendrick, the American historian who was writing a biography of Andrew Carnegie, all duly entered their names in the Skibo Guest Book during the decade of the 1920s. So that even if Skibo was no longer the sort of early British version of an Aspen Institute of high culture that Carnegie had tried to make it, it was seldom dull or commonplace.

Certain parts of the old pattern did not change, however—the piper in the morning, the organ music which Roswell himself occasionally provided, the Sunday evening hymn singing, and most certainly the fanfare that accompanied the twelfth of August, the opening day of the grouse season. This was now a more festive day then it had ever been in Andrew Carnegie's day. The great hall and the dining table would be bedecked with white heather for good luck, and the house would be filled with Roswell's hunting companions.

Except for the dark year of 1931 as the Western world slid into the abyss of the Great Depression, Louise Carnegie and the Millers did not miss a single summer at Skibo throughout the 1920s and 1930s. Louise would usually arrive first, sometime after having spent a week or so in London. On a few occasions, she went directly to Aultnagar for a week or two of relaxation before opening Skibo to the Millers and the visitors who would inevitably follow.

If she found that she needed more rest and that her walks were becoming shorter and slower in pace, she was still remarkably vigorous and healthy for a woman approaching seventy. There were, of course, the increasing number of deaths of old friends and relatives to remind her of her own mortality. Sister Stella died at Skibo in the summer of 1927, and her passing was as quiet and unobtrusive as her living had always been. But there were also births to lift Louise's spirits in the witnessing of the renewal of life. The Miller's third child, Barbara, was born on 3 April 1925, and made her first trans-

Atlantic crossing at the age of two months, not quite equaling the family record established by her older sister Dede.

After a busy winter in New York in which she continued active in her church work, as a trustee of the Carnegie Corporation and with various other philanthropic interests of her own, including the patronage of the fine arts, particularly music, Louise would look forward each spring with eagerness for the journey to Skibo. And if the years passed by in their even rhythm, so that it might have seemed difficult to distinguish one Skibo summer from another, there was almost always some signal event to differentiate one year from another: 1927—the summer that Louise Carnegie was presented to King George and Queen Mary at Holyrood Palace in Edinburgh; 1928—the year the family dedicated the Carnegie Birthplace Museum, next to the small weaver's cottage in Dunfermline where Andrew had been born ninety-three years before.

A summer to be remembered especially was that of 1926. So accustomed was Margaret now to the regularity of their annual summer treks that not even the fact that she was expecting another child within a few weeks could prevent her and her family from joining Louise at Skibo in early June. On 15 July, St. Swithin's day, the Miller's fourth and last child, Margaret Morrison Miller, or "Migs," as the family would later call her, was born, and the first child to be born in Skibo Castle since George Gray two hundred years earlier. Legend has long maintained that if it rains on St. Swithin's day, there will be rain on the next forty days, but if the sun shines on that day, there will be no rain for forty days. Louise Carnegie in happily recording in her diary that evening the birth of her last grandchild, failed to mention the weather, but for Skibo Castle and its occupants it was a bright day no matter what the published weather reports may have noted.

The year 1935 was a banner year for Louise Carnegie. There was a year-long celebration both in the United States and Britain of the centennial of Andrew Carnegie's birth. Louise had never sought the bright light of publicity, but in this year she was perforce shoved out onto the center of the stage. At Saint Andrews in June she was awarded an honorary Doctor of Laws degree—an unusual honor for a woman to receive from the ancient Scottish university which had twice elected her husband Rector. In September, just before she returned to New York, the City of Edinburgh gave her its highest honor, the Freedom of the City. Now she had her own scroll and fancy casket to add to that impressive collection of Freedoms that had been amassed by Andrew.

Back in New York, on Sunday, 24 November, she and the family gathered around the radio in her study to hear the BBC program broadcast directly from the small second-story bedroom in the cottage on Priory Lane, Dunfermline, where Carnegie had been born. The following day, the actual centennial date of Carnegie's birth, there was a special concert at Carnegie Hall, conducted by Walter Damrosch, which repeated the program of the opening night of Carnegie Hall, including Tschaikovsky's Fifth Symphony, which the composer himself had been present to conduct forty-four years earlier.

It must have been difficult not only for friends and family but for Louise herself to believe that it was her husband's centennial that was being celebrated—so hard to realize that it would soon be fifty years since they had married, over fifteen years since he had died. The years had gone by so very quickly, and now she herself was nearly eighty, but it seemed to her she was as much involved with living as she had ever been.

Much of her vitality and zest for life she could attribute to having Margaret's four children close by her during most of the year, playing in her gardens on 91st Street which she shared with the Miller household, and in the summer, turning Skibo into a child's garden of pleasure. Margaret did not make the mistake of insulating her own children from the world of childhood. Not only were the four close enough in age to have great fun playing together, but the gates of Skibo were open to the children of the area to come and romp with the Millers on the green lawns, play hide and seek among the rhododendron bushes, swim in the pool, and climb the great branches of the beech trees that obligingly swept low to the ground to receive youthful scalers. And Margaret and Roswell had their own friends, both from the immediate vicinity, such as Captain "Bob" Grant and his wife of Dornoch, and from England and America—sailing and hunting companions of their own age who brought their children to add to the general happy confusion. Both Louise Carnegie and old Skibo itself were engulfed in youth and seemed to grow younger in spirit as they quietly aged physically.

Among the new visitors to Skibo were Mrs. Emmeline Thomson of Edinburgh and her bachelor son, Gordon, a young lawyer in Edinburgh. Louise had met Mrs. Thomson many years before in Edinburgh, and meeting her again quite by chance at a Sutherland County agricultural show in August 1924, Louise discovered that Mrs. Thomson and her two grown children, Gordon and Edith, were on holiday in the Dornoch area. Louise invited them to come to Skibo for lunch. She and Mrs. Thomson thoroughly enjoyed each other's company, and Gordon, who was just Margaret's age, and the Millers immediately struck up a warm friendship. Roswell urged him to come back to Skibo the next summer for the salmon fishing and a real visit, an invitation which the young lawyer readily accepted. So began a close association with Gordon Thomson that would last for the remainder of their lives.

Thomson's visits to Skibo became a regular part of the summer schedule. Roswell found him to be a jolly good companion whether with rod or gun in the field or as a member of his crew on the yacht, *Wyndcrest II;* Gordon's skill as a conversationalist enlivened the dinner table talk for Louise and Margaret; and the Miller children adored him. He was the lively young uncle whom they did not have within their own family. He entered into their games with zest, never with that patronizing condescension that even very young children find insufferable in most adults.

The Thomsons lived at 26 Heriot Row in Edinburgh–on the same street and only a few doors away from where that master Scottish storyteller, Robert Louis Stevenson, had once lived as a child. It must have been the poems from the *Child's Garden of Verses* that Gordon recited to them as he pushed the Miller children in their swings, and surely it was that most thrilling of all childhood novels, *Treasure Island*, that he would read to them in the evenings. How exciting to have an uncle who lived in such close proximity to where their hero Stevenson had lived as a boy!

Young Louise–Dede–was particularly attracted to Gordon Thomson and early developed the same young schoolgirl's "crush" on him, the handsome, older man, that her contemporaries had for a favorite male teacher or a movie star. As the years passed and Dede matured into a beautiful and high-spirited young woman, her early affection for Gordon was not transferred to any young man of her own age. And as Dede approached seventeen and her senior year at Miss Spence's School, it also became apparent to the entire family that Gordon's feelings toward Dede were no longer avuncular but truly romantic. He was at last deeply in love.

Margaret and Roswell could hardly deny this love even though it must have seemed to them that Dede was terribly young to be considering marriage. Gordon was everything parents might wish to have in a man who sought to marry their daughter, a man well established who would obviously go far in his profession, and above all, a man who was kind and considerate and would certainly prove to be a loving husband and father. And what could Louise Carnegie say in opposition even if she had had a wish to do so? Although Dede must have seemed to her grandmother even more than to her parents ridiculously young to be considering marriage and to a man who was more than twice her age, still there was almost exactly the same difference in age between Dede and Gordon as there had been between her and Andrew. The only real difference was that her granddaughter and the man she had chosen sought to begin their life together when they both were some twelve years younger in age than she and Andrew had been when they were married. And that might be all to the good. So this new relationship with Gordon Thomson was accepted by the family. Plans were begun for a wedding at Skibo the following summer—July 1938—when Dede would have passed her eighteenth birthday.

The summer of 1938 was the busiest and most crowded with guests since the heyday of Andrew Carnegie's tenure at Skibo. After a three-week rest in London to gather up strength for what was to come, Louise Carnegie arrived at Skibo on 3 June to be followed by the Millers coming in at various times over the following week. The bride-to-be, after having spent ten days in Edinburgh with the Thomsons, arrived on 18 June with Gordon. And then the pre-nuptial festivities began as the Miller and Thomson clans and three of Dede's school friends as bridesmaids from New York all gathered in late July at Skibo.

At high noon on Wednesday, 27 July, Dede, wearing her mother's wedding veil, and Gordon, in his morning coat, were married in the Dornoch Cathedral. The hybrid ensign that flew over Skibo had taken on a new meaning.

After the wedding, there was a breakfast served at Skibo for a thousand guests, who ate their lobster and Scottish salmon and drank the eight hundred bottles of champagne to the heady skirl of the Dornoch Pipe Band. Then the bride and groom departed for the wedding trip abroad, but for the assembled guests, the festivities were not yet over. For the past three months, the crofters and farm laborers of the Dornoch and Creich parishes had been accumulating a gigantic pile of brushwood on the high ground above Overskibo farm. It promised to be the most impressive bonfire in the history of Sutherland County. When the summer sky finally grew dark enough around eleven in the evening, Margaret was asked to light the bonfire and the several thousand onlookers cheered. It was reminiscent of the nine bonfires that had been lighted above Cluny to announce Margaret's birth. Now the tenants and parishioners of another county of Scotland were celebrating a new generation and the hope that this day's event gave for the continuity of the Carnegie tenure of Skibo. The next morning, Louise Carnegie wrote in her diary, "We all went to bonfire at Overskibo, which Margaret lit. Near midnight when we returned. And so the great day is over. God bless them both."[14]

It was a fortnight before the last of the overseas wedding guests had departed and some semblance of normality could return to Skibo. And then the newlyweds returned from their honeymoon on 18 August, and for the first time Dede signed her name in the Guest Book not as Louise Carnegie Miller of New York, but Louise M. Thomson of 26 Heriot Row, Edinburgh.

The hunting season was not as festive this year as it had ben in previous years. The family had spent itself on the wedding activities. Moreover, in this late summer of 1938, there were other hunters out in the field with guns loaded, and they were after bigger game than Scottish grouse and red deer. Having bagged little Austria in March, Adolf Hitler now had Czechoslovakia within his sights, and in Britain there were hesitant and reluctant stirrings over the possibility of another war with Germany. There were, to be sure, or so it was generally thought, reasonable men at the head of both the British and French governments. Mr. Neville Chamberlain and M. Eduard Daladier had learned well the lessons of August 1914 when just a little patience, a little quiet and sane diplo-

View of Skibo taken from lower garden terrace. South side.

Laying of the cornerstone, new addition to Skibo, 1899. Andrew Carnegie to the right of the stone. Louise Carnegie holding Margaret to the left.

Margaret Carnegie Miller.

Steel engraving of Skibo, showing the hybrid flag of the Union Jack and Stars and Stripes sewn together.

macy would have prevented the quite unnecessary tragedy of a world war in which there had been no victors, but many ruined losers. Only that notorious jingoist, Winston Churchill, who had long been in exile from British politics, was crass enough to point out that 1938 was not 1914 and that Adolf Hitler was not Kaiser Wilhelm.

Reassuring as the Chamberlain government might be, conditions were nevertheless unsettled enough for Roswell to advise an earlier departure than usual from Skibo that year. The Millers, without Dede of course, and Louise Carnegie arrived back in New York in time to hear with thankfulness that although Hitler had been allowed to get the prize he sought without firing a gun, the important thing was no guns had been fired on any side. Mr. Chamberlain and sanity had prevailed. "I believe it is peace in our time . . . peace with honor," the British prime minister told the cheering crowds at Heston airport on his return from Munich on 30 September. And isolationist America also cheered. The Carnegie-Miller family, listening to the BBC announcement over the radios in New York, believed Chamberlain. They were already looking forward to next summer when they would be reunited with Dede and Gordon at Skibo.

Louise Carnegie, coming first as was her custom, arrived aboard the *Aquitania* at Southampton on 8 June 1939. Dede, radiant with the happiness of her first year of marriage, was at the dock to meet her. They spent three days in London together and then four days at 26 Heriot Row in Edinburgh before Louise headed on north to Skibo. The Millers arrived ten days later bringing Dede with them, and they all settled in to enjoy another summer at Skibo.

But it was a curious summer—this summer of 1939, which proved to be Louise Carnegie's last at Skibo. Few visitors came from America in those last months of uneasy peace and Britons stayed at home, close to their radios, for by now it was apparent to all that the "our time" for which Chamberlain had promised peace was going to be of very short duration. Having first tasted the succulent Sudetenland given to him at Munich and then having gorged himself upon the whole of Czechoslovakia in March, Hitler had once again gone out hunting for even bigger game. Now Poland was the target. On 23 August the news was flashed to the world that the impossible had happened–Hitler and Stalin had signed a non-aggression pact giving each other *carte blanche* to divide up the carcass of still another ally of Britain and France. Now it was no longer a question of if but rather how soon the war would come. It was imperative for the Millers and Louise Carnegie to make plans immediately for their return to the United States. Until return passage could be arranged, there were many things to be done at Skibo–blackout curtains to be hung at all the windows, the entrance lights to be painted a dark blue so they could not be seen from the air, and Louise's prized lemon verbena plants that grew close to the castle foundation to be dug up so that sand bags could be piled around the ground floor windows.

The last visitor to Skibo for that summer was Louise Carnegie's good friend, Margaret Buckworth from Weymouth, England. It was agreed that Louise would return to Weymouth with her while Roswell went to London to see what accommodations he could get to return them all to New York. Margaret and the three younger Miller children would see to the closing of Skibo and then join her mother and Roswell as soon as passage was arranged.

On the morning of 1 September, Louise Carnegie bade her last farewell to Skibo. The servants were lined up in the front hall, as they did each fall, for Louise to shake their hands, and say goodbye. Andrew's old valet, the ever-faithful Morrison, who was still in their service, helped Louise to try on her gas mask to see if it fit properly. Then she walked out to the car where her new factor, Mr. Whittet, was waiting to accompany her to the railroad station at Bonar Bridge. And there in the driveway stood Grant, the piper, who began playing the plaintive "Happy we've been a' t' gither" and "Will ye no come back again."

It was all so horribly unreal, and yet so horribly familiar, like a nightmare that recurs and one knows in advance how terrible it is all going to be. It was just twenty-five years almost to the day that she and Andrew had left Skibo for what proved to be his last time. Now she was older by three years than Andrew had been in 1914, and she was more of a realist than Andrew had ever been. She must have known that surely this was her last moment at Skibo, the "fairyland of peace" that was no more. Weeks after she had left Skibo, one of the few servants who remained in the castle, found tucked behind a picture of the family upon one of the tables in the drawing room, a small bouquet of flowers, now sadly wilted, with a note in Louise's handwriting attached, "Farewell to Skibo. September 1, 1939."[15]

Two days later Great Britain and France declared war on Germany, and the nightmare of world war had begun again. Roswell was fortunate to book passage on the *Nieuw Amsterdam*, belonging to the Holland-American line. The trip across the Atlantic would be less hazardous and more pleasant than the Carnegies' flight from war had been in 1914 for this was a ship of a neutral country and there would be no blackout and little danger of a torpedo attack. But the departure was even crueler this time for they were leaving behind one of their own, and no one could say when, if ever, it would be possible to see Dede and Gordon again. Before departing, Louise found time to do some shopping in London for baby clothes as the Thomsons were expecting a child in January. So in actuality, they were leaving three of their family behind to endure this new, and what promised to be, more terrible war.

Back in New York, the same old waiting game that Louise knew so well had begun all over again. But this time, it was quite a different war, one that initially gave promise of not being a long war of attrition in the trenches. Within a month, Poland had been savagely torn in two by the German panzer divisions driving in from the west and the Russian armies swarming in from the east. Poland's two allies, Britain and France, could only sit helplessly on the sidelines. Their formal declarations of war against the German Reich were only that—declarations on paper without substance on the battle field. Poland, as a nation, had quickly joined Czechoslovakia and Austria, another lost shade sunk in oblivion. Like a sudden, furious summer storm, the lightning had struck, and then it was over.

While Nazi and Soviet gorged on their defenseless prey, all was indeed quiet on the western front. The French hunkered down in false security behind their Maginot Line, while the British waited in their blacked-out island for what this unconventional war might bring to them from out of the sky or from across the North Sea. Perhaps it would be nothing. Perhaps Hitler, with the absorption of Poland, had finally been sated. But even though some London wits spoke jokingly of Britain's new "Bore War," no one really believed that this conflict could be so quickly resolved, for except for Poland, nothing had been resolved. Hitler's troops could still roam at will on the darkling plain that was Europe, and nowhere was there a secure shelter should the blitzkrieg strike again.

Louise Carnegie and the Millers found the first winter of suspenseful waiting almost impossible to endure. Letters from Edinburgh were meant to be reassuring. Yes, Dede was feeling fine during these late months of her pregnancy; yes, they had enough food even though it was rationed; yes, excellent medical care was obtainable and a good nurse had been secured for the time when her services would be needed.

But for Margaret Miller, three thousand miles away, letters were not enough. She suddenly decided that she must take advantage of this unexpected hiatus in the storm of war and go over to Scotland to see for herself how matters stood and to be with Dede when she gave birth. Roswell offered no objections. "Good idea. Why not?" were his words of blessing on the trip. As for Louise Carnegie, however, such stout encouragement could not be so easily or directly given. Margaret told her mother of her decision as they sat together one evening just before Christmas listening to carols being sung over the radio. After a long pause, all Louise could manage to say was, "You've a long trek to *your* manger."[16]

It was one thing to get the support of her husband and mother for this potentially hazardous journey. It was quite another matter to get the consent of her government for a trip to a belligerent zone by a neutral on what would have to be considered a non-essential mission. It finally took the intervention of the Carnegies' close friend and senior partner in the J. P. Morgan Company, Russell Leffingwell, to get directly from the Secretary of State, Cordell Hull, the special passport visa and documentation necessary for Margaret to make the trip.

With Mary Cheyne, who was eager to visit relatives in Scotland, as a traveling companion, Margaret sailed from New York aboard the Italian liner *Conte di Savoia* on Wednesday, 27 December 1939. From the age of three months, Margaret had made many trans-Atlantic trips, but this was her first crossing in mid-winter, and she was taking this in the midst of a great war. Italy was still technically neutral, and the ship gave pointed emphasis to this important fact by running with its lights blazing and with a bright spotlight focused on the Italian flag painted on its side.

Margaret had the good sense to keep a diary of this adventure and the literary style to give a graphic portrayal of what it was like for a middle-aged American woman to travel by ship across the Atlantic and then by train and plane from Genoa through France to Britain in that winter of 1940. The ship seemed spookily empty—only thirty-two passengers in all of the first-class quarters. On New Year's Eve, alone in her cabin, Margaret opened the silver flask her mother had given her and quietly drank a toast "to the grandchild I hope 1940 is going to bring to Roswell & me!"[17]

After a short stop in Naples, the ship docked in Genoa on 4 January, and from there, the two women went by overnight train to Paris. The City of Lights still had its lights on. The crêpes at La Rue and the lobster thermidor at Prunier were as sinfully rich and as easily available as they had been the last time Margaret was there. The war—even the rumors of war—had as yet changed nothing for the Parisians. Only the plain glass windows in Ste. Chappelle gave testimony of some apprehension for the future.

The American embassy, however, seemed far more conscious of the true state of affairs. Margaret was not given a warm welcome in that official mansion at 2 rue Gabriele when she appeared to request permission to fly on a British plane to London. The embassy officials were not particularly impressed that the purpose of this trip was to visit a daughter in Edinburgh. "The first man I saw wanted me to have a letter sent from my daughter saying she earnestly requested my presence with her at this time!" Margaret later wrote in her diary, "I nearly laughed aloud picturing Dede composing such an epistle. I said, 'She doesn't know I'm coming. You couldn't expect me to tell her till I got there!' " But if the officials were not impressed with her reason for traveling by air on a plane of a belligerent nation, they were impressed by her credentials—her passport personally signed by Cordell Hull, and her letter of introduction from a Morgan partner. Finally, Margaret recorded, she wrote on the form, " 'My daughter is an expectant mother' (hateful phrase) '& earnestly requests my presence with her. It is necessary that I proceed at once. Any other mode of travel to England' (they would like to send you by the Hook of Holland so that you could take a neutral boat. . . . so America would not feel responsible for your life if this boat were flying a Dutch flag) 'would take too long.' Well, I got the permit stamped on my passport."[18]

On Monday, 8 January, she and Mary Cheyne flew from Paris to London and checked in at the Connaught.

> They were distinctly surprised to see us at the Connaught, as they were told no Americans were allowed to cross over! . . . At 4:45 I put in a call for Gordon & it came through at once. How many times in the last ten days I have pictured myself in the telephone booth in the hall where I last heard Dede's voice in September and heard myself saying, "Gordon, this is Mummie." Well, I did just that & there was complete silence at the other end. I repeated "This is Mummie Miller at the Connaught Hotel" then there was a gasp & he said the word "Dede." They must have been together having tea for then she broke in, the same darling, old Dede & never did anything sound so sweet as her voice! We didn't talk long for tomorrow the floodgates can be opened. But she is fine & it was no mistake to come over —a thousand times no! She sounds so well & happy & think of actually talking with her again. That's enough for one day, I can't take in the thought that we'll be seeing each other tomorrow.[19]

The Flying Scotsman left the King's Cross Station at ten a.m. the next morning precisely on its old familiar schedule just as it would continue to do throughout all the years of war. It is Britain's crack express train from London to Edinburgh, but it could not fly fast enough to satisfy Margaret. Not one minute late, the train pulled into Waverley Station, and there were Gordon and Dede, shouting, "Mother," as the train came to a stop. "We went right up to the pleasant room they have engaged for me [in the North British Station Hotel] and hurled questions & answers back & forth at each other. It certainly was a complete surprise to them. Only last week Dede had said, 'if it hadn't been for the war Mummie would have been sailing!' "

Paris might not be taking the wartime blackout seriously, but the cities of Britain were. "It was a dense fog outside, the worst they have had all winter. We tried to get a taxi to go to their house, but not one was running in Edinburgh! So we linked arms, Gordon in the middle & walked along Prince's Street to take a tram. Then I saw a *real* blackout. You just can't imagine anything blacker than Prince's Street in a fog without lights."[20]

The following days passed quickly. Waiting was not difficult in Edinburgh, with shopping errands in the morning, an occasional movie in the afternoon, "cozy teas" in the library with Dede and Gordon, a few games of backgammon or chess in the evening after dinner, and then the taxi ride back to the hotel through streets lighted only by a quarter moon shining over the black silhouette of the castle.

The war was never far from their minds. Occasionally, they could hear gunfire out over the North Sea as RAF planes drove off the occasional German raiders, seeking to disrupt shipping into the Firth. On one afternoon, Margaret and Dede walked to a near-by cinema to see Basil Rathbone and Douglas Fairbanks, Jr., in *The Sun Never Sets*. "A young sailor & his girl came in & sat right in front of us. His black hair was clipped short & well-oiled & his neck above the square blue linen collar was pink & smooth—that of a boy! All at once I saw many bodies like his floating in the North Sea after the sinking of the 'Courageous' & the 'Royal Oak.' "[21] For the young men who shipped out into cold seas, where twenty minutes in the water meant death, or for those who flew missions of coverage above the ships in this January of 1940, there was little humor in quips about a Bore War.

One week after Margaret Miller arrived in Edinburgh, on 16 January, her first grandchild, Elizabeth Carnegie Thomson was born. "When I went up [to see Dede] she was just her old self & looking fine as a fiddle. The wee one was wrapped in shawls lying in the Moses basket. I couldn't see much but one big blue eye! Well, I left then & drove back through the white carpeted streets full of very, very happy thoughts."[22]

On 1 February, Margaret took the long train ride north from Edinburgh to Bonar Bridge. This was no "Flying Scotsman" express, and the train seemed to crawl north through Dunfermline, Perth, Killicrankie Pass to Inverness, where Margaret changed trains. After Inverness, the little local going on up to Bonar Bridge stopped at every village. "I could see dark stations where the only light was the yellow, round eye of the flashlight held by the guard. Out of that blackness his voice came calling the names we know so well & as we reached each station I became more & more excited." At Edderton, the train had to wait for twenty minutes for a train headed south. "My heart was beating fast. . . . Sitting there knowing Skibo was just across the firth was a hard test of patience. . . . Finally we pulled into Bonar Bridge. . . . There was one bright star shining over the firth as we crossed the bridge & came onto Skibo. It was quite a different home coming, but the oak wood in the vestibule smelt just the same & the fire in the hall made it home."[23]

Margaret woke early the next morning. "I thought the light would never come. . . . From six o'clock I waited for it, so impatient I couldn't sleep. At last I heard the cry of a curlew, then the wild turkeys in the fields below the house & there seemed to be a beginning of light so I got up & looked out of the window facing south. A few flakes of snow were falling. . . . The field below the house was ploughed & lying in even brown furrows, the fountain was empty & the lawn a yellowish green. . . . The sand bags which had been filled the first days of the war were just as they had been placed."

Margaret had never seen Skibo in the winter, but she found

that in this season too it had its own peculiar charm. She took a long walk after breakfast, down the golf course road, past Loch Evelix, and up the Monks Walk. She found the snowdrops and for the first time saw them in bloom. She was enchanted by their beauty. In the pine woods, "two roe deer ran away. It seemed to me their coats were very dark—& two woodcock got up from the bushes at the head of the mid-way . . . I crossed the bridge & walked through the arches made by the bare, gray branches of the beech trees, & down the long aisle of the Monks Walk enjoying the increased view ahead of one given by the leafless trees. I came back *loving* this new Skibo with her sails well reefed & her ropes taught [*sic*] to meet the winter & the storm of war. Somehow she seems dearer & more real than ever before with the fields ploughed to raise food crops, her size cut down to provide comfort & security for our family while the tempest rages; none of all her pomp & luxury . . . but yet with a reality & a permanence of lovableness about her I have never felt before."[24]

It was a strange week of mixing the familiar with the unfamiliar, and it passed all too quickly. Margaret rashly squandered nearly half of one month's four-gallon ration of petrol to take the family's little red Ford on a tour of the entire estate: past Spinningdale over to Lairg and the falls of the Shin, then stopping to take pictures of snow-covered Aultnagar, and across the empty, brown moors back to Skibo. On other days, she called on old neighbors and friends in Dornoch and visited tenants and employees to inquire after their sons who had gone off to war. On Sunday, she attended services at the Dornoch Cathedral and was pleasantly surprised to find that the old cathedral, always so cool and damp in summer, was now quite comfortably warm.

Margaret was back in Edinburgh on 7 February in ample time to help with the plans for her granddaughter's christening at St. George's Church, scheduled for 18 February. And then there were less than two weeks left of her visit—days filled with lunches, visits with old friends and with new friends of Dede and Gordon, and mostly just being with her daughter, knitting and talking, and watching little Betty splashing in her bath, nursing, and developing rapidly into a healthy Scottish bairn.

That zealous old conservator of Margaret's infancy and youth, her Nana, came down from Dumfries, where she lived in retirement with her sister, to see her former charge. It was not an easy reunion for Margaret. She was shocked to see how greatly Nana had failed both mentally and physically. Time past and time present had become intermixed, "and the most painful [are when] a ray of memory breaks through to see her going about my hotel room putting my shoes together & straightening things up. Almost broke my heart." Only when Margaret took her up to Dede's home did the Nana of old, in complete command of any situation, reassert herself. She held little Betty in her arms with professional assurance and said to the baby, "Oh how I wish I was going to take care of you!"[25]

Margaret's last day in Edinburgh with the Thomsons was spent as quietly and normally as possible. A long walk, with Dede and Betty in her new pram, down Prince's Street and up on the Mound below the Castle where they met Gordon and then back to Heriot Row. "Dede handed Betty to me after her six o'clock supper & left us alone for one last cuddle. Then we played some new records on the Victrola. We managed to get through the last hours together creditably & even to enjoy them. Gordon was a great help for he talked very interestingly at dinner. I remember he told us about the celebrations in London after the Armistice was signed." And then too soon the taxi was at the door. One last glimpse of "Dede's dear head outlined against the dimly lit hall behind the open door. . . . From the direction of the North Sea four search lights converged in one glowing spot of light. All around was complete blackness."[26]

Margaret had the bad luck of having her diary confiscated by the wartime censor at the Heston airport, but fortunately, after it had been carefully scrutinized, it was returned to her in New York many weeks later. Otherwise, the return trip to Genoa, where she and Mary Cheyne were to board the Italian liner *Rex*, was quite uneventful. There was even time while in Paris to rent a car like any peacetime tourist, and drive down to Fountainbleau, stopping for lunch in a charming little *relais* at Barbizon. It might have been the bright summer of 1930 instead of the dark winter of 1940 except for the long stream of army trucks, camouflaged with paint, that passed them on the road back to Paris.

The *Rex* sailed from Genoa on 5 March and nine days later docked in New York. Roswell, Barbara, and Migs were on the pier to greet her. "When things quieted down a bit I ran over to see mother. Such a lot to talk about and such eager questions about Dede and Betty. The family came to tea and all admired the pictures of the new member of the family. I shall always be grateful that I could go over and spend those happy weeks with Dede."[27] So ended Margaret Miller's wartime diary.

Margaret had come back to the security of the United States with little time to spare. Six weeks after her return, the wartime recess abruptly ended as the blitzkrieg struck again, first unexpectedly, in Denmark and Norway, and then in May in Holland, Luxembourg, and Belgium. The Lowlands were captured in a matter of days, and then the German troops poured across the Belgian border into France. The Maginot Line was useless—a fence protecting the front door while the Germans came in the unguarded back door.

Now the waiting for the future became more agonizingly painful than ever before—waiting to see if Britain would be invaded and could possibly survive now that Hitler had turned all of western continental Europe into a Nazi fortress. This time there was no talk in the Carnegie household, as there had been in 1915, about keeping America neutral and arranging peace talks with the warring powers. There should be no further negotiations with Hitler, and both Louise and the Millers ardently hoped that the United States would soon enter the conflict and attempt to save what was left of Western civilization.

That hope was fulfilled in December 1941. Roswell was called back into the navy as a lieutenant commander, and because of his first-hand experience of the coastal waters around Scotland, he was sent to the Firth of Clyde as naval adviser. Robin, a freshman at Princeton, was on an accelerated course which would allow him to be commissioned in the navy within two years. Margaret herself became a Red Cross worker and an airplane spotter, while her two younger girls also gave up further schooling to help in the war effort, Barbara, as a nurse's aide at Bellevue Hospital, and Migs, now sixteen, as a driver for the Red Cross—"my truck-driving granddaughter," was Louise's proud boast.

Roswell on weekend passes got to see the Thomsons on occasions. They even opened Skibo for a ten-day Christmas party for some of Roswell's fellow officers and the Thomsons' friends in dark December of 1941 immediately following Pearl Harbor. For a brief moment old Skibo, behind its blackout curtains, had glowed with something approaching its customary cheer. This visit inspired Dede and her father to turn Skibo into a convalescent hostel for British and American service men. But the estate was too remote and transportation at too great a premium to make that wartime use of Skibo a reality.

Louise Carnegie, now eighty-five, gave what strength she had to the promotion of the Bundles for Britain organization. "I'll keep in close touch with Bundles for Britain," she wrote

Dede, "my fingers are too stiff to sew or knit but I can still write cheques, so I'll help them with contributions for whatever Mrs. Latham tells me they need most."[28] It lifted her spirits somewhat to think that out of these thousands of bundles perhaps one might find its way to Dede's canteen.

Everything, even nightmares and wars, must finally come to an end, and Louise Carnegie hung on to life waiting for the peace that must follow. The tide slowly turned—D-Day, June 1944, and the Allies were at last back in France; V-E Day, May 1945, and Germany surrendered; finally, six years almost to the day after Britain declared war on Germany, Japan surrendered aboard the *U.S.S. Missouri*, and it was all over.

Foolish as she knew it was, Louise began to think of Skibo for next summer. Skibo's magic was powerful, but not powerful enough to enable an eighty-nine-year-old woman to return to its charms. But Margaret and Migs in March 1946 flew to Scotland, and shortly afterwards, Dede and little Betty, now six years old, flew to New York. Betty was introduced to Great-Grandmother Naigie. It brought back to Louise Carnegie memories of her taking baby Margaret to see her great-grandmother nearly fifty years before. The circle was now complete and it was time to depart.

On 24 June 1946, Louise Carnegie died at her home in New York. The era of Andrew and Louise Carnegie was at an end, but Louise had the satisfaction of knowing that Skibo would not be abandoned. Margaret had built the place into her life as completely as Louise ever had. And now there were three generations that Louise was leaving behind to enjoy her beloved Skibo.

VI

Skibo, en famille: *The Margaret Carnegie Miller Years* 1946–1980

Skibo now belonged to Margaret Carnegie Miller to have and to hold for what would prove to be a longer period of time than it was the possession of either her father or her mother. In some respects, to be sure, her mother's will had merely made legal what had always been true. Skibo had really belonged to Margaret throughout her life. It had been her birth that had prompted Andrew Carnegie to find and purchase the estate, and it had been she, at the age of two, who had given the final pat of the trowel to the laying of the cornerstone of the new addition that made Skibo the imposing castle that it is. It was her willingness to make Skibo a part of her adult life that had encouraged and enabled her mother to continue to hold Skibo in stewardship for her. It had been at Skibo that her youngest child had been born, and it was here that her oldest child had met the Scotsman whom she would marry and give to Margaret her five Scottish grandchildren. Skibo, more than anywhere else was the one place that had given a continuity to Margaret's life. And now it was to Skibo she must return in the late summer of 1946, only a few weeks after her mother's funeral, to claim that which was truly her birthright.

She and Roswell flew from New York on the American Overseas Air Lines, the line for which her son-in-law, Lennart G. Thorell, daughter Barbara's husband, was a pilot. Arriving by train from London, the Millers were met in Edinburgh by Dede and Gordon and their four little daughters, who then accompanied them on up to Skibo. Once again there was the familiar drive from Bonar Bridge to the gates of Skibo, now decorated with a huge sign, WELCOME TO SKIBO. Margaret Miller was home again, but this time she was not returning as the laird's child. Now she was herself the Lady of Skibo. She suddenly realized as they drove through that avenue of trees past Loch Ospisdale that all of the staff and many of the tenants of Skibo would be at the castle door to greet her. All of the painful shyness she had once known as an invalid child seized her. "What ever shall I say when I arrive and the employees of Skibo are there to welcome me?" she whispered in consternation to Dede. "Keep calm," the always poised Dede reassured her mother. "You know very well that you will say the right words when you get there." And so she did.

Mr. Whittet, the factor whom her mother had employed prior to the war to replace the aging Hardie, was there to introduce her to the staff, the faces and names of many being new to her. And then in the pause that followed with everyone looking at her, Margaret said quite simply, "Thank you so very much for coming to welcome me. I know we will get on very well together." The staff cheered their assent. Margaret's was not the flowery oration her father had given upon his arrival in 1898 to accept the lairdship of Skibo, or the graciously regal acknowledgment that her mother had bestowed upon her employees in 1920, but Dede had been quite correct. Margaret's words were her own right words, and in their ingenuousness and sincerity, they struck the keynote of her tenure as Skibo's rightful owner.[1]

The six years of World War II had brought far more

changes to the farms and policies of Skibo than the four years of World War I had. The golf course had been turned into a grain field, for besieged Britain had needed every possible acre of arable land to help feed its inhabitants even the very limited rations they received. With a shortage of labor and materials, proper maintenance of the estate buildings had been impossible. There were roofings and painting to be tended to and cracks in the swimming pool to be repaired. Some of the tenants' sons who had been in the service were returning to their fathers' lands, and they would be welcomed back, but new tenants must also be found. The lands of Skibo must once again become as productive as possible for food was still at a premium and strictly rationed in postwar Britain. This was not only a civic duty, Whittet informed Skibo's new owner, but also one of practical necessity for her as well. Andrew Carnegie had never had to concern himself with how much it had cost to operate Skibo. It had been his playground, and playthings are not expected to turn a profit. Even Louise Carnegie, in spite of her talk about reducing expenses and living more simply than the Carnegies had in the past, had never really had to worry about Skibo's being a revenue producing proposition. But now with property taxes in Britain at an all time high and with less than half of the fortune of her mother left to her after having paid the inheritance taxes, Margaret Miller would have to make Skibo pay its own way. This would require careful and thrifty management, and if there was one factor in all of Britain who was up to that task it was undoubtedly Whittet. Louise Carnegie had chosen far more wisely for the future than she could have imagined at the time she first employed him.

For Margaret, the readjustment to a more simple style of living at Skibo presented no great hardship. Even more than her mother, she found enjoyment in simplicity rather than in grandeur. Her father had had to push hard—inordinately hard, as he once said—to escape from the poverty of his childhood, but her problem had been to keep from being surfeited by luxury. Margaret would not have restored the old glory of Skibo with its road walkers, its manicured golf greens, and its lavish dinner parties even if it were still financially feasible to do so.

Life at Skibo, to be sure, would never be what anyone could call Spartan. If the domestic staff was reduced from eighty-five to less than thirty, with five gardeners instead of eighteen, neither her family nor her houseguests would suffer from a lack of personal service. Many of the old amenities would be kept. If there was no professional organist to give them Bach and Wagner while they dined, so much the better. Now Roswell, Dede, and Barbara could play away to their hearts' content. And of course, there was still Hugh Grant, the piper, whose matutinal piping was for Margaret not a luxury but a necessity.

The gamekeeper was equally indispensable for the family, and in Harry Blythe, Skibo had a fine gamekeeper and an expert instructor in teaching the science of stalking the Royal stag, of raising a covey of partridge or grouse, and in casting for the black-backed salmon. Blythe's father and grandfather before him had been chief gamekeeper for the Montrose family, and young Harry as a lad had learned every fine detail, every nuance of field and stream sports as they had been developed in Scotland over the centuries.

Scottish hunting may lack the polished formality of the English fox hunt (a sport which one wit described as "the exquisitely ineffable in hot pursuit of the absolutely inedible"), but it is far more purposeful in its objective, for the Scottish hunter pursued what is definitely and deliciously edible. And hunting in Scotland does have its own protocol and ritual which make it vastly different from the American frontier practice of going out into the field with any gun that might be available and taking a quick pot shot at anything that moves.[2] Scottish hunting requires all of the order and finesse of an army moving in perfect discipline to execute a highly complicated maneuver. And the gamekeeper is the commanding general, responsible for deploying his troops of stalkers or beaters, ponies, and of course the gentleman and lady hunters themselves to the most advantageous position in order to gain the objective of bagging the game. The success or failure of the mission ultimately depends upon him, and his command must be obeyed without question, not only by the menials but also by the gentry, no matter how exalted their rank. One must understand and work with nature in order to best it. Even the greenest novice in the field must respect immediately the sacred commandments of the hunt: first, in driving game, such as grouse or partridge, the hunters or guns, must stay in a straight line and fire straight ahead as the birds are driven into the air by the beaters–never, ever turn to the side or rear to shoot if the birds should fly overhead and behind you; second, in stalking deer, never shoot until the stalker gives the word to shoot–only he will know when a shot is feasible; and third, shoot to kill, for if a stag is only wounded, he must be pursued, no matter how far he may lead you, and finished off. While there is a stiff penalty of a £100 fine for anyone found leaving a deer carcass behind after the hunt, no gentleman and true sportsman, regardless of the fine, would ever leave an animal to die slowly and agonizingly of a wound that the hunter has inflicted. These are cardinal rules and anyone not observing them will never again be asked to join a company of respectable hunters.

In stalking a red deer, one must follow the stalker wherever the hunt may lead–crawling on all fours through the brush if necessary, or crouching in absolute immobility for long intervals if commanded by a hand gesture to do so. For nature has given to the deer the keenest senses of hearing and smell of any of its creatures. A carelessly snapped twig can send a grazing animal into instant flight a half a mile away, and if the wind is in the right direction, a deer can detect a human being two miles away. Since the rifle has a range of only 150 yards, one must close in on the target with a greater stealth and deftness than that practiced by even the most skilled of pickpockets.

Assuming that one's shot is a good and true shot and the stag is felled, then the ritual of dressing the kill must be performed in the field with the solemnity and respect the noble beast deserves. The animal is bled and the carcass is gralloched or gutted. When this is done the gamekeeper turns to the successful hunter and asks with great formality, "Is this, sir, your first stag?" If the answer is yes, the gamekeeper dips his hand into the stag's blood and smears the hunter's face on cheeks and forehead with the blood of his kill. These smears must remain on the hunter's face until after the evening dinner. They are his proud badges of honorable initiation into the hunt. The Shetland pony, equipped with a deer saddle, is then called up. The deer is slung across and secured on the saddle. Then the whisky flask is brought out and to the cry of "Slàinte!," a toast is drunk.

The greatest prize of deer stalking is, of course, the Royal stag, that magnificent buck who has twelve points or tines to his antlers. Favored indeed of Diana is the hunter who at least once in his life has this monarch of the high moors within his gunsight and can then salute his noble fallen prey with the appropriate Gaelic "Slàinte."

Except for some of the small common game such as rabbits and pigeons which are always available for hunting, each game has its own open season, firmly established and rigidly enforced throughout Britain. The setting of these seasons is not the capricious act of some bureaucratic commission but rather has been determined by the individual pattern of each species' natural life rhythms. Salmon fishing extends from 15 January until the last day of September, but it is at its best in early spring when the first salmon begin their relentlessly frenetic run from the sea up the rivers and streams to spawn.

The trout season does not begin until the middle of March when the ice leaves the streams and the hungry trout are ready for the expertly cast fly. The twelfth of August, the opening season for the grouse, is perhaps the most publicized opening day of all, and in the Scottish Highlands it ranks along with New Year's day and Robby Burns's birthday as a glorious day of celebration.

The season for red deer stalking does not have as precise a beginning or an end as do many of the other game seasons. A stag is ready to be hunted only after he has cleared his new antlers of their early protective covering of velvet. The red deer sheds his last season's antlers in the early spring and immediately begins sprouting new antlers which will be fully mature by the mating season in the fall. The stag ignores the hinds throughout the summer. His main preoccupation besides grazing is in clearing his antlers of their velvet by rubbing them against trees and rocks, and this task is usually accomplished by the first of August. Then, and only then, can the stag be hunted. The season is relatively short—only through the months of August and September, for by the first of October, the stag's attention is diverted from his belly and his regal head adornment to collecting a harem of hinds with whom to mate. His neck begins to swell, his eyes become bloodshot and the high moors resound with his bellows of defiance to other males who would dare to invade his territory.

With the onset of the rutting season, another sports season comes to an end. The gentry store their rods and guns and depart from the Highlands. But, for the gamekeeper and his assistants, the most difficult part of their year now begins, for each hunting estate in Scotland is assigned its quota of hinds that must be killed in order that the deer population may be kept under control for the future health and conservation of the species. Hind hunting in the cold snows of November, December, and January is not the festive sport of stag stalking in the bright sunny days of September. It is cheerless and often bitter drudgery and requires infinitely more skill and perseverance, for the hind is much smarter and more adept in protecting her own and her yearling calf's life than is the stag. But the quota must be filled, and the sooner it is accomplished the happier both the gamekeeper and the remaining hinds will be. By February, there is at last on the high slopes a winter quiet, unbroken by the crack of a rifle. In June, the hinds drop their calves, and the whole cycle begins again for both man and beast.

All of these facts of life pertaining to the field and stream Harry Blythe taught the children and grandchildren of Margaret Miller, and in Robin and Migs he found his most enthusiastic and able disciples. For Robin, in particular, hunting and fishing were the two most important pursuits of his life. And Harry Blythe, living in retirement in Inverness many years later, would remember Robin Miller as being the finest sportsman he had ever known, without exception.[3] Coming from Blythe, this was high praise indeed.

In the summer of 1947, Margaret Miller wished to delay her trip to Scotland in order to be with Barbara, who was expecting her first child in August. So it was arranged that Roswell would go over earlier and she would join the family late in August. Their son Robin, who had not been able to get across the previous summer and consequently had not seen Skibo in eight years, was also eager to get there in time for the twelfth of August season opening and to introduce his wife, Ann Brinton, whom he had married in 1945 while still in the Navy, to the delights of his favorite spot on earth. Robin and Ann arrived at Skibo on the evening of 10 August, where they were greeted by Roswell and by the Thomson family who had come up a fortnight earlier to open Skibo for the summer.

Robin, on Christmas seven years earlier, had been given a handsome hunting log book by his mother, and for the next thirty-five years he was to record meticulously in that journal the events of every day in which he was fortunate enough to be able to pursue his favorite sports of hunting and fishing. On the evening of 11 August, he duly noted, "Back at last. . . . Very sweet to be back. . . . This morning the Thomsons and I fished the Evelix for a couple of hours, . . . the stream was extremely low. . . . All we caught were some small trout."[4]

Robin seldom made any entry into his log book that was not directly related to either hunting or fishing. Certainly he would have thought it unnecessary to record that his sister, Dede Thomson, after their first fishing try of the season, mentioned that evening that she was not feeling very well. Perhaps after wading around in the stream all morning and with the unusually hot weather in the afternoon, she had caught a slight summer cold.

The next morning, Dede felt worse and so, much to her disappointment, she had to miss the opening day of the grouse season. It proved to be not a good day for the hunters. The moorlands were dusty dry and it was unbearably hot out on those open, unshaded fields. The hunters returned with only a small bag of grouse, and Robin entered into his log book the disgruntled comment, "The 'Grand and Glorious' Twelfth was very disappointing this year, . . . the grouse so scattered and the moors so dry the dogs couldn't point decently."[5] But the disappointment the Millers felt over the poor opening day on the moors was quickly forgotten in their growing concern for Dede. She was now running a fairly high temperature and complained of her whole body aching. It must be a touch of influenza—nothing to be alarmed about really, as long as she was kept as comfortable as possible and isolated from her five young children.

But the next morning, the 13th of August, the paralysis struck swiftly and mercilessly and she was unable to move any part of her body. The hastily summoned physician gave the dreaded diagnosis–poliomyelitis. This was a rare disease in Scotland, and in provincial Dornoch there were no adequate facilities for treatment. It soon became apparent that there would not be time to move her to Edinburgh. The paralysis quickly affected her respiratory system. Her breathing became more labored and at nine that evening, Louise Thomson died. She was twenty-seven years old, the mother of five children ranging in age from Betty, who was seven years old, to their only son, William Gordon, who was seven months. Hers had been a short but crowded and wonderfully happy life, and now, incredibly, it was over.

It had all happened too fast to comprehend. Just two days earlier she had been in the best of health, happily casting her fishing line into the Evelix, and then suddenly there had been no time to inform Margaret back in New York that her daughter was ill. No time even for her mother to get to Scotland for the hastily arranged private funeral service. It was not until the 28th of August that Margaret had sufficiently recovered from the shock to fly to Britain and join her grief-stricken family at Skibo. No one could understand where or how Dede might have contracted the disease. But it was pointless to speculate about a futile question like that, for now the living had to be concerned with the living. Gordon Thomson especially knew the heavy responsibility he faced. At the age of fifty, he had become the single parent for five small children. It was a challenge which in the years ahead he would meet with admirable skill and great success. He would never consider another marriage. He had had but one great love in his life, and that would sustain him for the long years that lay ahead.

But things would never be quite the same again for any of them. Gordon Thomson would always consider the death of his wife to have been a turning point in the family's history. Not only was life at Skibo to be different, but relationships within the family changed in ways that they might not have done had Dede lived.[6] Roswell Miller, in particular, was not the same person. He had always been particularly close to his first child, an attachment that, if anything, had been strengthened by her marriage, for her husband was Roswell's boon companion, and Louise's and Gordon's children were his first grandchildren and his especial delight. Now in his grief, he seemed inconsolable. He stayed on at Skibo a few weeks after the funeral, and his son Robin tried to distract him with hunting and fishing and short sailing trips on the yacht, but he only went through the expected motions in a desultory fashion. Finally, after Margaret arrived and could assume the role of head of the family, it all proved too much for him. He fled from Skibo and went off to France alone where he could be away from the family and from the painful association of memories that life at Skibo held for him.

Over the next few years, although he and Margaret continued their annual visit to Skibo, it was apparent to the family that there was a difference in their relationship. Margaret remained steadfast in her old attachments, but Roswell seemed distracted, no longer as interested in the familiar patterns of their life that sustained his wife. He sought and found new interests.

In 1953, Margaret and Roswell Miller were divorced—quietly and with the minimum of rancor that could be expected in the termination of a marriage of such long duration as theirs had been. Roswell kept the estate in Millbrook, which they had named Migdale, after one of their favorite spots at Skibo. Their home on 90th Street in New York was given to the Carnegie Corporation of New York, to whom Louise Carnegie had earlier bequeathed her own mansion. The Corporation, in turn, leased both properties to Columbia University to serve as a graduate center for the New York School of Social Work. For a number of years, Margaret did keep the old carriage house, also on 90th Street, as an apartment where she stayed occasionally when in the city. Now more than ever she regarded Skibo as her true home.

Without Roswell upon whom she had previously depended to assist her in questions involving Skibo and her other properties, Margaret Carnegie Miller now had to make the decisions regarding the management of her estate. And it was she who must maintain the Carnegie family's participation in the various philanthropic trusts which her father had established. She was grateful to have these responsibilities, and above all she was thankful for Skibo. It provided a continuing interest to her life, and it was the single lodestar that kept her expanding family with their diversified interests on at least one commonly shared course. No matter where her children and grandchildren might be, she knew as summer approached, their thoughts like hers would be directed toward Skibo even if they weren't able to get there every year.

Margaret was also eternally thankful for the Thomson family. The five children all had their mother's physical beauty, personal charm, and that wonderful capacity for enjoying the good things of life without being spoiled by all that had been given them. When she was in their company, she could not feel she had really lost Dede. And Gordon Thomson remained a steadfast and true support to her life. Although her equal in years, he was really a part of her children's generation. He could thus be both another son to her and at the same time, her contemporary–a wise counsellor and true friend. Gordon, in his chosen profession of the law, had brought great distinction to the family, and Margaret could enjoy vicariously his many accomplishments. After having been a King's Counsel in Scotland, in 1953 he was appointed a Senator of Her Majesty's College of Justice in Scotland and took the Judicial Title of Lord Migdale, which pleased Margaret immensely.

In the years that followed Dede's death, Skibo became

more than before a second home to the Thomson family. It was they who would go up to Sutherland early in May each year to prepare for the opening of Skibo to the rest of the family in the summer. The Thomson children spent a great part of every summer there, and so it is hardly surprising that Margaret felt she knew them better than any of her other grandchildren. She also felt a special responsibility toward the Thomson children, who had been left without a mother at such young ages. It was Margaret's practice, as it had been Louise Carnegie's before her, to come to Britain each spring by way of London, and spend a few days there prior to going on up to Scotland. During the years in which Margaret and Louise Thomson were attending a girls' school outside of London, Margaret Miller would always visit her granddaughters at their school, and take them out to lunch or dinner, or perhaps to the theatre in London. As their father could seldom be away during the school year from his court responsibilities in Edinburgh, their grandmother was a much appreciated surrogate parent, whose visits they eagerly awaited each spring.[7]

It was memories of summers at Skibo with their grandmother that all of the Thomson children would especially treasure, however. Betty would remember as a very small child sitting on her grandmother's lap and being fascinated with the look and touch of the amber pendant which Margaret would take care to wear for her granddaughter's enjoyment. And William would remember his grandmother's understanding and forbearance toward some of his youthful pranks. One Sunday, while all of the rest of the family went off to church, he stayed behind to entertain the young son of the Countess of Sutherland, who was visiting him. Entertain his guest, Willie certainly did. The young Scottish lord was eager to try his hand at driving a car, so Willie obtained the keys to one of the family's autos. They started off bravely enough, but unfortunately his friend could not keep the vehicle on the gravelled drive. The auto plowed across the neat green lawns of Skibo, leaving deep tracks in the turf, softened by recent rains. Fortunately, the novice driver was able to bring the car to a stop before it plunged off into the terraces below or crashed into a tree, but the two boys waited in great trepidation for the church-goers to return home. Margaret took one look at the evidence of the short-lived joy ride, but said not a word. They all went silently into lunch, and still not a word was said. After lunch, the two boys slunk embarrassedly outside only to find that the gardeners during the lunch hour had swiftly and efficiently erased the shameful tracks. But the lesson had been learned. His grandmother's silent correction of his misdoing had been far more effective than any angry scolding could ever have been.[8]

The older Thomson girls, Betty, Margaret, and Louise, would remember another adventure of their childhood that had a sadder ending. One day, the girls discovered in one of the garages a magnificent vehicle that looked as if it could have come straight out of a Charles Dickens novel. It was their great-grandfather's famous four-in-hand coach, undoubtedly the one Andrew Carnegie had used when he set out in 1897 to find a Highland home for his family and had discovered Skibo. The girls were enchanted with it and persuaded the ever-obliging chauffeur, Edgar, to have it wheeled out on the grounds so that they could play in and on it. Even the two younger children, Mary and Willie, entered into the fun, climbing up to take turns in the driver's seat and in the seats on top, just like the pictures they had seen of their Great-Grandfather Naigie's coaching rides seventy years before.

Then the children thought of a more exciting game. They persuaded Edgar to hitch the coach to a farm truck and slowly drive it around the driveway while Willie sat in the driver's box holding the reins. Another servant recorded the event with a movie camera, carefully angling the shots so that the source of locomotion was not revealed. When their grandmother later saw the film, it appeared as if Willie were actually driving a four-in-hand coach pulled by horses. No one disillusioned her, and in consternation, Margaret Miller ordered the coach burned. When she later understood what had actually happened, she gave in to the children's entreaties and rescinded her order. But the word came too late. Andrew Carnegie's historic coach had already been reduced to ashes.[9]

But there were many more happy than unhappy memories of childhood at Skibo. There were the wonderful picnics in the Fairy Glen, the rides across the moors when the heather was at its richest purple and the swimming parties in the pool. There was that glorious day for the ten-year-old Willie when Harry Blythe judged him to be old enough to handle a gun properly and took him out in the fields to teach him how to shoot. Skibo was as much a fairyland of delight for the Thomson children as it had been for the Miller children a generation before.

It was also a place of wonder for Margaret's American grandchildren even though they did not have the opportunity of enjoying it every summer as did their Scottish cousins. Robin and Ann had two sons: Roswell IV, also called Robin, and Kenneth. After Robin's divorce from Ann in 1955, he married Helen Johnson by whom he had a daughter, Dorothy Louise. Margaret's second daughter, Barbara, who was married to Lennart Thorell, had four children, Linda, Carolyn, Lennart, Jr., and Sandra; and the Millers' youngest child, Migs, had two children, Barbara and Gail, by her first husband, and another daughter, Pamela, by her second husband, Robert Finlay. These ten grandchildren also had their individual memories of the magic that was Skibo.

Robin and Kenneth Miller, like their cousin Willie Thomson, felt they were being initiated into manhood when their father and that super-hero of their childhood, Harry Blythe, took them out into the field with their own guns, or taught them how to tie that special lure, the Harry Blythe fly. Barbara's oldest daughter, Linda, was five when she first visited Skibo. She would later write, "I do remember outings such as floundering when the tide came in (or went out?), when Grandma and Grandpa [Miller] led a procession of family and friends through the mudflats—everyone armed with spears. I remember the marvelous fruit—strawberries the size of plums, raspberries and gooseberries like huge grapes—fruit with a flavor that might have come from the Garden of Eden. I remember Grant the Piper, piping around the castle every morning and then I remember watching the adults through the banister above the big hall as they linked arms and two by two filed into dinner, as Grant piped them to the dining room."[10]

Later, like her great-grandmother Louise Carnegie before her, she began to keep a diary of her visits to Skibo. One entry, made when she was twenty, told of "a lovely peaceful tea by the rushing stream. The sun shone brightly and we [her Grandmother Miller and she] sat talking as dopey bumblebees sucked nectar from the Scabious and then promptly flew into our teacups." Margaret told her granddaughter of the old days when she and her mother and sometimes her father would come up to the cabin in Fairy Glen by themselves, with no servants about and here Louise Carnegie taught Margaret how to cook and Andrew Carnegie would enthrall his daughter of how back in the ancient days of the Picts and the Scots, long before even the Romans had come to Britain, the fairies danced here by moonlight and left their dance circles forever marked by a more verdant turf. And all of this held as much wonder for Linda as it had for Margaret sixty years before.[11]

Linda would also recall the summer of 1967 when there

was an abundant crop of chanterelle mushrooms in the woods of Skibo. As she would later write of this event: "These are a wild orange mushroom with an upturned cap which looks anything but edible, but have the most delectable flavor." Margaret Miller's good friend, Robert Gordon Wasson, a senior partner in J. P. Morgan and Company and one of the world's greatest amateur mycologists, had given Margaret a book on mushrooms. "Grandma's book on mushrooms assured the 'hunter' they [chanterelles] could be mistaken for nothing poisonous. Even so, I think we were quite foolish. Well, with my inexperienced youth and Grandma's bravery, the two of us had the cook sauté some. Grandma shared a half with me one night. We survived the night and so the next evening each ate some more. Still surviving, we decided they were quite safe and so delicious we would indulge in as many as we wanted. I think Grandma with her spirit of adventure is the only one who would have braved such an experiment with me."[12] Quite obviously, Margaret Miller gave to her grandchildren more than just material wealth, and it was her zest for living that they would treasure.

Her granddaughter in retrospect acknowledges this heritage. "Skibo to me was for all its splendour, a place to enjoy the simple pleasures in life. There was a time to collect wild flowers, pick brambles, take walks and enjoy the beauty of the earth we live on. It was a place where one could be quiet and have time for thought. I experienced its quieter years, when life was more mellow. To enter that magnificent home, the smell of peat clinging to its very fibers, and curl up in front of the glowing peat embers in the fireplace and listen to stories of earlier times, was part of a beautiful heritage I am so lucky to have had."[13]

Linda Hills was quite right. She and her siblings and cousins would know Skibo only in its quieter years. The old pageantry was gone, the tempo of its life had slowed, but mellowness and quiet have their own charms. There once had been a time for fireworks, but now was the time for firelight, and each in its own good season illuminated Skibo with its appropriately lovely light.

Not even the Twelfth of August was quite as festive a social occasion as it had been when Roswell Miller had filled the castle with hunting companions, but Harry Blythe would stay on until old age forced his retirement in 1965, teaching each new addition to the family the proper decorum for young gentlemen—and young ladies, too—in the field and stream. Migs and Robin had the same exuberant love for rod and gun that their father had, even though they didn't need a large company to celebrate the opening of the seasons with them. Each got to Skibo every summer or fall that it was possible to do so, and Robin continued to record scrupulously in his hunting log each day's bag or catch. Although not a man ordinarily given to fancy words either in speech or on paper, Robin could become almost poetically effusive in recording a particularly fine moment of hunting or fishing. This was his entry for 14 September 1947, which was the final entry for what had been an exceptionally poor season because of the heat and dryness.

> The last day at Skibo, and what a day. . . . Fished Evelix this morning with no luck at all. . . . In the afternoon we hoped to have a last try at the pigeons, but the weather was so completely foul that fishing seemed a better bet so Dad, Mother and I went back to Evelix. . . . Wind was fairly light from the west but during the afternoon increased until the loch was completely swept and the rain squalls poured down like great, gray sheets. At last a real break in the weather but just too late! About three o'clock, before the wind had risen too much between the rain storms, I had the most exciting fifteen seconds of the whole summer. It was

The great entry hall of Skibo.

The Morning Room. Portrait of Mrs. Carnegie above the fireplace.

The Formal Dining Room. Portraits of Carnegie's heroes on the wall.

The Library of Skibo.

what I had been waiting eight years for . . . . to take a big fish with a light tackle. . . .

When we were drifting up the north shore of the lake just below the lowest green grassy point well off the bank, [we cast out.] . . . The next cast brought no results; the following one, a great black back rolled over. . . . It was one of those moments when all time and action seem to stop. You could see the shining head below the water, the dark, iridescent purple of his back, and the ribbed dorsal fin, jagged on top. For seconds he hung there, poised for the strike, then the line went taut and the rod arched. I struck and felt the hook sink—God!—"Got him!" Then all hell broke loose! He ran away from the boat at 45 degrees for twenty-five to thirty feet, then cleared the water in a beautiful leap which showed me a fish of some seven or eight pounds. When he landed he was going at right angle in the completely opposite direction. The line was slack. The second jump showed he was still on and our hopes rose momentarily, but our hopes were premature, for another run and jump proved fatal. In this third jump, he went straight up in the air for his whole length, shaking his head savagely and then down he came on his back—and *off!* The pre-war gut just couldn't take the punishment and it is doubtful if any leader of that kind could have.

So, I didn't get a salmon in the Evelix this summer, but I have a memory which is a challenge to try again real soon.[14]

Hugh Grant, who had served the Carnegie Miller family for many years as its piper and who also assisted Blythe in the fishing and hunting expeditions, died suddenly of a heart attack in November 1960. Robin in his hunting log made one of his few personal entries that was not directly related to sport. On 6 November 1960, he wrote, "I have lost one of my closest and dearest friends. . . . [Grant's] very sudden death from a heart attack on November 3rd was a real shock to all of us and felt to me like the ending of a chapter in my life. He was one of the very finest gentlemen I shall ever be privileged to call my friend."[15]

Grant's death also marked an end of a chapter in the life of Skibo. Margaret Miller did not attempt to replace him with another piper, and for the first time since 1898, there was no skirl of the pipes in the morning to arouse the family and their guests to the day's activities. Another bit of Skibo's colorful pageantry was gone.

The great organ in the hall was also silent. After sixty years of service, it needed major and expensive repairs if it was ever to be played again and since it had been so seldom used in recent years, Margaret felt it was not worth the cost. The exuberant sounds of Skibo so gaudily offered up in Andrew Carnegie's day were now a part of the past. Along with so much else, they were now important memories to relate to the youngest grandchildren but no longer necessary to duplicate. For Margaret, these small economies were a part of the valiant effort she was making to bring the costs of Skibo somewhat closer in line with the revenue it produced.

As part of the new economy, Margaret Miller at last carried through in the plan for disposing of that part of the Skibo estate which Louise Carnegie had discussed some forty years earlier with John Ross and then had so quickly abandoned. The building of a large hydro-electric plant on the Shin River which, it was generally believed, would destroy the excellent salmon fishing on the Shin, helped reconcile the family to the loss of this part of the estate. It meant giving up the magnificent waterfall, which had been Andrew Carnegie's pride, and it meant losing Aultnagar, which meant much more to Margaret. But Aultnagar was really no longer needed. Skibo Castle itself had become an Aultnagar—the family needed no additional retreat to find peace and quiet by themselves. Aultnagar, now under new ownership, became an inn, open to the general public, who could walk its forests, climb its hills, and marvel at the superb views of the Shin valley.

Mr. Whittet, the factor, was Margaret Miller's staunch ally in all of these economy moves. For thirty years, he gave to the estate his expert managerial skill, doing his best to increase the revenue and to lower the costs. He often regretted that Andrew Carnegie had been as determined and as successful in his philanthropy as he had been. If Carnegie had only left a sizable endowment specifically designated for the maintenance of his Highland home, Whittet's task would have been an easier one.[16]

Although they worked closely together in attempting to make Skibo self-supporting, Whittet was not successful in obtaining Mrs. Miller's consent to raise the rents on the farms. These rents had been set decades earlier and now with the inflation that postwar Britain had experienced, they were, in his opinion, ridiculously low. But Margaret had not abandoned her childhood concern for those less fortunate than she, an attitude which had so greatly amused her father but which her factor found to be unnecessarily generous.

By the early 1960s Whittet had grown old and weary of the struggle and he began to give serious thought to retirement, which Margaret did not discourage. One morning in April 1964, Robin Miller, at his home in Salt Lake City, Utah, was surprised to receive a call from his mother. She was leaving for Scotland. Could he possibly join her at Skibo as quickly as possible? She was retiring Mr. Whittet and would greatly appreciate her son's help in interviewing candidates for the position of factor. Robin obligingly secured the first available air transportation to Scotland and by the next evening was at Skibo.[17] He dutifully went through the interviewing process with her, but in the end, Margaret Miller made her own choice. She selected J. M. McLeish to be the new factor.

The appointment of McLeish proved to be a highly controversial event in the history of Skibo. The old employees never accepted him. They had found Whittet to be a hard taskmaster but a fair and equitable supervisor, one who expected "an honest day's work for an honest day's pay." That he had been the best factor in all Scotland was the general consensus in the villages on the estate. It would have been difficult for anyone to step in and live up to Whittet's reputation. Loyal old time employees like Bobby Edgar, who had been chief mechanic in charge of the family yacht, and his wife, who had been the cook on the yacht, remembered with nostalgia the glory days of the 1930s when the Miller family was a happy unit, sailing the *Wyndcrest II* in the coastal waters and firths of Scotland. The present under McLeish seemed bleak in comparison.[18]

Margaret Miller's family and her lawyers in Edinburgh and New York, on the other hand, were far more charitable in their judgment of McLeish as factor. They felt that under the circumstances, he had given Mrs. Miller the service she had desired. She had not wanted or expected expensive maintenance. The days of yachts and large staffs of employees were over. The castle itself and its policies were as beautifully maintained as ever, given the limitations of staff.[19] Moreover, McLeish found himself to be always in a difficult position, as indeed Whittet had been before him. Mrs. Miller, unlike her mother, was a most accessible person. Any employee felt free to go directly to her, where he or she could find a receptive and sympathetic audience. Such an easy and democratic regime made for warm and kindly relations between mistress and employee, but it did not make the task of supervision an easy one for any factor.[20]

If there were tensions and dissension beneath the surface between her factor and her employees and tenants, however, Margaret Miller seemed to be happily unaware of them. As the years passed, she and her beloved Skibo seem to be gently aging in beautiful harmony together. Slow decline is natural for places as well as people, and Skibo, without golf courses, yachts, hordes of employees, even without a piper, still suited her very well. If anything, Skibo Castle seemed to grow ever more beautiful with the passage of time. It no longer gave that appearance of newness that had made it stand out in stark contrast to its natural setting. It now looked weathered by time as a castle should, and its white pink sandstone had become softened by a delicate patina. Here Margaret had her family, her few long faithful servants, and the beauty of her flowerbeds and the good food from her own gardens, greenhouses, and farms. She asked for little more.

Skibo had fewer guests in these years. Members of the family generally were quite enough. There were, of course, a few old friends from Edinburgh and London and even from New York who were warmly received to Skibo's hospitality. Trustees from the Carnegie Dunfermline Trust and the United Kingdom Trust made their occasional summer visits, and Margaret was always delighted to have news of how the various philanthropic agencies were carrying out her father's vision of wealth being used for the benefit of all mankind.

Alan Pifer, the president of the Carnegie Corporation of New York, Andrew Carnegie's largest philanthropic foundation, and his family were particularly welcome visitors in 1979. The Pifers were her neighbors in Southport, Connecticut, where Margaret had her winter home, and it was grand to be able now to show them her Highland home. Margaret was particularly delighted to discover that Mrs. Pifer's given name was Erica, the Latin name for heather. When they were getting ready to depart, Margaret ordered the staff to do something that had not been done for many years. She told the staff to put up above the gate, "Will you no come back again," the old Carnegie valedictory to departing distinguished visitors. Margaret, in turn, was to receive her own gracious acknowledgment of appreciation. After the Pifers had gone, one of the maids discovered on the bed occupied by their young son, Daniel, a piece of paper upon which he had printed "THANK YOU." It meant more to Margaret than any obligatory formal note of thanks from a prime minister or a duke.

There were a few rare occasions in these years, to be sure, when old Skibo would put on her formal dress and manners to reveal herself to be still the Grande Dame that she had always been. Never to be forgotten was the afternoon in June 1964 when the Queen and Prince Philip came for lunch. Lord Migdale who would later purchase the ancient estate of Ospisdale, had been appointed Lord Lieutenant Commander of Sutherland in 1962. As the Queen and the Prince were making an official visit that summer to mainland Scotland's northernmost county, it was Lord Migdale's responsibility to welcome Her Majesty to Sutherland, and Skibo was the obvious place to entertain the royal party for lunch. So the invitations went out and were accepted.

On the afternoon of 25 June 1964, Margaret stood at the front door and welcomed her guests to Skibo. Migs had arrived from the States to assist Margaret in the entertaining, and of course Lord Migdale was on hand to give the official greetings of Sutherland. This was no impromptu, informal affair as the sudden arrival of King Edward VII for tea had been sixty years earlier, and Margaret had somewhat nervously awaited the day for many long weeks. But everything went off with grace and style. After lunch, the Queen followed Margaret into the drawing room for coffee. She walked immediately over to the floor length bay window and looking out across the meadows to Dornock Firth and beyond that to the city of Tain and the distant Ross hills, she said, "It is so beautiful, I don't wonder you love this place." Elizabeth and Philip then signed the Guest Book, taking a whole page for their royal signatures, and as they got into their car which would take them back to the royal yacht *Britannia*, the Queen said, "We look forward to seeing you this evening." That evening Margaret, Lord Migdale, and Migs were rowed out into Cromarty Firth where the *Britannia* was anchored. It was a beautiful evening, and the yacht was festive with its colored lights, gaily patterned deck chairs, and a small orchestra on the fantail softly playing dinner music. The party was served supper on the deck, and then after an hour or so of gentle pleasantries exchanged between the Queen and her guests, the orchestra struck up the anthem, "God Save the Queen," to signal that the party was over. It had been quite a remarkable day for both Skibo and its mistress.[21]

Although the Fourth of July Fête Day, and the Children's Day had long ago been abandoned, Margaret in these last years did carry on one practice begun by her mother many years before of opening Skibo to the general public. On one Thursday afternoon in July and again for one day in August the Castle grounds and gardens would be opened for the benefit of the Sutherland Women's Rural Institute. A small entry fee would be charged and in the late afternoon a brace of grouse and a salmon from the Skibo larders, choice grapes from the greenhouse, vegetables from the garden, and bottles of whisky would be raffled off. All proceeds were donated to the Institute and on a good day as much as £300 would be taken in as both local denizens and visiting tourists took advantage of this opportunity to view the fabled castle and perhaps win a prize of its produce.[22]

Margaret Miller was a staunch supporter of the local Dornoch Pipe Band, and particularly after Skibo no longer had its own piper, she would several times during the summer go into Dornoch on Saturday evenings to hear the band's regular weekly performance. In recognition of her patronage, the secretary of the organization, William Grant, in the summer of 1980, Margaret's last year at Skibo, asked permission to return the favor of her coming to Dornoch to hear the band by bringing the band out to Skibo. Margaret Miller would never forget that wonderful evening of 11 June when the Pipe

Band, in formal Highland dress, made its appearance in the castle drive. As she would later write, "After they had played out on the lawn, they came into the front hall and gave us a half-hour none of us will ever forget. Before ending, the secretary came up and asked what tune we would like to have. My daughter Migs asked for 'Amazing Grace' and I asked for 'Scotland the Brave,' ending with 'Happy we've bin a'togither.' "[23] So for one brief time the halls of Skibo resounded again with the exuberant skirl and flourish of the pipes so often heard in the past, and it sounded very good indeed.

Margaret had long been a familiar figure to the people of Dornoch and Bonar Bridge, as she visited the local hairdressers, did small shopping errands, attended services in the cathedral on Sunday mornings and spoke easily and informally with the trades-people and old neighbors and friends on the street. But unlike her father, she had never enjoyed and usually tried to avoid the formal public appearances which her position in the community offered her. She did serve each year as judge for the children's flower arrangements at the annual Women's Rural Institute festival, and she could be counted on to be present for the Sutherland agricultural fair and to be in the seat reserved for honored guests at the Highland Games, but she preferred not to have the bright spotlight of publicity focused on her.

In 1977, however, she agreed to present the Carnegie shield to that year's winner of the annual Dornoch Royal Golf Club Open. Dornoch boasted of having one of the oldest and one of the ten finest golf courses in the world. In 1901, soon after taking up residency at Skibo, Andrew Carnegie had given to the Royal Golf Club a large handsome sterling silver shield, embossed with two views of Dornoch Cathedral, appropriately enough since the first golfers in Dornoch had been the ministers of the Cathedral. The shield also bore the inscription, "This shield was presented by Mr. and Mrs. Andrew Carnegie of Skibo Castle to the Dornoch Golf Club

MCMI." It is surely one of the most impressively valuable golf trophies in existence, and on 25 August 1977, Margaret Miller with a simple little statement of congratulations presented the shield to the winner, Kenneth Houston, a youth from near-by Golspie. Houston's acceptance was equally charmingly unaffected, "I could never have won if my mother hadn't managed to bring me up a hot meal," was his modest rejoinder.[24]

There were also moments of excitement at Skibo in these last years to break the quiet and even tenor of its routine. One early summer day, in 1970, Magnus Magnuson, a director of an independent film company, wrote Mrs. Miller asking for permission to come to Skibo to take pictures on location for a documentary film the company was planning to produce on the life of Andrew Carnegie, to be called "The Star-Spangled Scotsman." For two weeks, a team of actors, directors, camera and prop men were in Dornoch, taking pictures of the actor who was portraying Carnegie as he fished on Loch Evelix, and of the Carnegies' taking tea in the gardens. They even brought from retirement old Angus Macpherson, the Carnegies' first piper at Skibo, now over ninety but still able to play the pipes. Once again Macpherson made his familiar circle around the castle grounds while the cameras ground away. On another summer, Margaret gave permission to the Sutherland Conservative Association to use the grounds below the terraces for a garden fête. A huge tent was erected for the party's benefit sale and for speeches to fire up the faithful for the coming elections.[25]

There were other events in these years, to be sure, that were not as pleasant for Margaret and must have evoked some bittersweet memories. Roswell Miller in 1970 returned to Skibo for a visit, and for the next several summers continued to make a brief appearance. The old magical attraction of Skibo after all these years had proved to be still irresistible. Margaret welcomed him to Skibo in the same manner as she welcomed other old friends of her youth who sought and rejoiced in being able to experience anew a Skibo revisited. Margaret accepted this situation with a grace and dignity which did honor to the standard her mother had set as a hostess.[26]

Margaret was to know a sadness at this time, however, that was very painful to accept. In the spring of 1975, she lost her only son, Robin. He and his wife had been visiting friends in the Bahamas in the late winter of that year. The last entry in his log book, dated 28 February 1975, gave an account of a rather unsatisfactory fishing expedition off the Grand Cays. Returning to the cold dampness of Connecticut in March, Robin caught a cold which developed into pneumonia. On 5 April, he died in an Essex hospital. The following summer his ashes were brought back to Skibo and scattered on the moors that he had so often happily tramped in search of the elusive grouse. In the little Creich parish cemetery near Bonar Bridge, Margaret had a stone marker erected by the north wall, facing the marker for Dede along the opposite wall. It read, "Roswell Miller III 1922–1975. He loved the heather and the skirl of the pipes."

Fortunately Gordon Thomson, who had retired from the high court justiciary in 1973, and two of his daughters, Margaret and Mary, were now permanently in residence at Ospisdale, and as her nearest neighbors, were close by to give Margaret their much needed support. Young Margaret was an enthusiastic farmer, and her prize Clydesdale horses usually took the highest honors at the annual agricultural show. And there was also Migs, almost as ardent in her love of fishing and hunting at Skibo as her brother Robin had been. Migs lived in Colorado, but each summer her mother, after arriving at Skibo, would send Migs a cable which always read the same, "Tight lines and hot barrels." This was her code to her daughter to inform her that the salmon were running and the grouse were on the moors. And if at all possible, Migs would fly over, eager to feel that line grow taut and see those guns blaze hot under an August sky at Skibo.

By her mid-seventies, Margaret Miller began to suffer from arthritis, and as the affliction became progressively worse, her physical activity became ever more limited. It became a painful trial now to take even a short walk down the drive past the ancient and gloriously blooming rhododendrons, but it was a test of her strength and endurance which she insisted upon each day. She became increasingly dependent upon a companion, who would accompany her to Skibo each summer, to write her letters—for it was an effort now just to sign her name, and to do small shopping errands for her.

Now more than before, Skibo became a place of quiet repose, a place to sit on the terrace and enjoy the pastoral views. In the summer of 1975, Margaret invited her friend, Ruth Adams, a former librarian in neighboring Westport, to come over to Skibo for a two-week visit. Miss Adams frequently helped with Margaret's correspondence at her home in Connecticut, and Margaret was eager for her to see at last the glories of her other home. Ruth Adams kept a careful record of her visit that summer, noting the menu for each meal, the short trips they took by auto to Ullapool on the west coast, and to Dunrobin Castle and Tain close by, and the visitors that came to Skibo during her short stay. She found the daily menus to be varied and delicious, with most of the food—the lamb, the salmon, the venison and beef, as well as the vegetables and fruit—being produced or caught on the estate. Miss Adams's notes give a detailed picture of the daily routine at Skibo during these last years of Margaret Miller's tenure.[27]

Another perceptive observer who had the opportunity of experiencing and recording over much longer periods of time Margaret Miller's last years at Skibo was Reathel Odum of

Washington, D.C. Early in 1977, Margaret had to find someone as a replacement for her companion of the previous summer, Alice Hoskins.[28] Mrs. Hoskins suggested her friend, Reathel Odum, who had had a distinguished career, first as John W. Snyder's secretary when he was a banker in St. Louis and later when he had gone to the Office of the Comptroller in Washington in 1931, to supervise the receivership of the many banks that were failing in these years. When Snyder's old friend, Harry Truman, arrived in Washington as the junior Senator from Missouri in 1934, Snyder agreed to Truman's employing Miss Odum as his secretary. For the next twenty years, she would be closely associated with the Truman family, first as his secretary during his ten years as Senator and Vice President, and then upon Harry Truman's sudden elevation to the Presidency, as Mrs. Truman's secretary and Margaret Truman's companion on her concert tours. With these credentials, Reathel Odum was highly qualified to serve as a secretarial companion for Margaret Miller for the summer. When the position was offered to her in February, she accepted. The prospect of spending several months in a castle in Scotland was most attractive.

Because of her worsening physical condition, Margaret Miller had by this time also found it necessary to employ a full-time nurse, Daphne Kunkler, a native of Jamaica. On 1 June, the small party of three left Kennedy airport on British Airways for Scotland. After spending a night at the Caledonian Hotel in Edinburgh, they were driven north to Skibo by Pocock, the chauffeur. It was after seven when they arrived at the castle, but the sun was still high in the sky, and Skibo was even more beautiful than Reathel had imagined it could possibly be. She was given as her bedroom, Louise Carnegie's old upstairs sitting room, which was next to Margaret Miller's own room. The opulence of Skibo was quite overwhelming even for someone who had once had an office in the White House. Even though the style of living at Skibo was now much more informal and simple than it had been in the Carnegie days, there was still an observed protocol of relationship with staff and tenants that an American from the Midwest often found confusing and somewhat difficult to adapt to.

In her frequent letters to the Snyder family back home and in the diary she kept for the summers 1977 and 1978, Reathel Odum gives a graphic account of life at Skibo during these last summers of Margaret Miller's residency.[29]

The daily routine seldom varied: breakfast promptly at eight-thirty at a small table placed in the bay window of the dining room; mornings spent in dealing with correspondence, reading the newspapers and discussing the day's activities; sherry at one; lunch at one-thirty; then generally a two-hour rest for Margaret unless a motor trip had been planned; tea at four-thirty before a peat fire in the hall; drinks in the hall at seven; dinner at seven-thirty; finally an hour or two reading aloud, preferably biographies or histories, or watching television until it was time for Margaret to retire. There were occasional short motor trips to the west coast of Scotland, or along the North Sea to Brora, or the familiar circle route from Bonar Bridge north past Aultnagar to Lairg and then across the moors to Loch Fleet and back to Skibo.[30]

So the days passed—quietly and serenely, without many notable variations to distinguish one from another. If often the days seemed long to Reathel and if occasionally she felt homesick for the busy and independent life she knew in Washington, she nevertheless found Skibo a most interesting experience. A real attachment developed between Margaret Miller and Reathel Odum which was more meaningful to each of them than a simple employer-employee relationship. And Skibo itself was the same place of enchantment and delight for Reathel as it had been for the many other visitors and residents who had preceded her. It was an exciting experience to sit in the hostess's position, which Margaret had asked her to do, and pour tea for Lady Elgin, Lady Bruce, and the Countess of Sutherland. Visitors were few in these years but they were still noteworthy even by Washington's inner circle standards. Reathel Odum felt she learned a great deal from Margaret Miller, her family, and her guests. It was a different world which few Americans ever have the opportunity to view, much less be a part of. Reathel made some very good friends during these summers–the Reverend James Simpson, minister of the Dornoch Cathedral, and his wife; the Jack McLeods; and the women at the Sutherland Women's Rural Institute.

Margaret took considerable pride in the way in which her most recent companion had adapted herself not only to the life at Skibo, but to the community as well. She congratulated Reathel on her handling of the raffle on the first Thursday afternoon that the gardens were opened, and she was delighted when Reathel was asked to make the formal presentation and preside at the annual festival of the Institute.

Margaret's family provided most of the social activity of Skibo in these years, however. There were frequent lunches and dinners at Ospisdale, and the high point of each week for Margaret was Sunday morning when she and Reathel would stop at Ospisdale to take Lord Migdale with them to the Dornoch Cathedral to hear one of James Simpson's fine sermons. Migs came to Skibo each summer and Barbara every other summer. There were also the fifteen grandchildren, and then a whole new generation of great-grandchildren began to make its appearance. Margaret no longer felt equal to having, nor did she feel that her staff should be expected to have, such large influxes of young people and babies at Skibo, but fortunately Ospisdale was close by, where the grandchildren and great-grandchildren could stay, and the young ones were always welcome to come over and swim in the pool and romp on the lawns as their parents and grandparents before them had done. And it was a wonderful break in Skibo's usual quietness to hear the wild splashings in the pool and the excited squeals of laughter and crying in the castle grounds.

As the summer of 1977 drew to a close, Margaret Miller asked Reathel if she would return next year as her companion. The Thomsons also urged her to do so, and she promised she would. Somewhat to her own surprise, she found that Skibo had also become a part of her life.

On 16 May 1978, Margaret Miller and Daphne met Reathel at Kennedy airport, ready to return to Scotland. This time they would be flying in the new supersonic Concorde jet, which had only recently, against considerable opposition, been allowed entry into the New York area. It was an exciting experience for all three of them—three hours and twenty-five minutes from Kennedy, New York to Heathrow, England at a speed of fourteen hundred miles per hour. Margaret must have recalled the stories her father had told her of his first crossing at the age of twelve from Scotland to New York, which had taken six weeks. It seemed as remote in time as the Stone Age. In spite of the speed of their crossing, however, they still had to spend the night in London before getting another plane to Inverness. When they finally reached Skibo on the evening of 17 May, all three were exhausted. Margaret called the trip "an unforgettable experience" and one she wouldn't have wanted to miss—but once was enough. It was too difficult to board and deplane twice. The spirit was more than willing but the body had become too reluctant. Never again would they come to Skibo via London, she firmly announced.[31]

Some changes had occurred at Skibo during their nine months' absence. The head gardener, Sterling, had left to take another position, and one of Margaret's first tasks upon her return was to interview candidates for that position. Some of the tenants in the near-by cottages had also moved on, including the family with the distinguished name of Sutherland, whose young children Reathel had made friends with the previous summer.

But what was surprising was not how much but how little things changed at Skibo from year to year. It seemed, in its quiet pastoral beauty, as if it were isolated from the ravages of time and change, and one could easily imagine that life would continue to flow gently along the same well-worn groove on into infinity.

But of course it couldn't, and Margaret Miller was made painfully aware by her own increasingly limited physical activity that she must begin to consider what should be done about Skibo when she would no longer be able to make the annual trip across the Atlantic. She discussed the matter with her family, particularly with Gordon Thomson and her daughter Migs. Lord Migdale counselled that she should carry out what Margaret felt would be the wishes of her parents—that when she could no longer enjoy Skibo she should give it to one of the Carnegie trusts or some other community service institution for the benefit and enlightenment of society, just as her mother had given the mansion on East 91st Street to the Carnegie Corporation of New York.

Migs, on the other hand, could not bear the thought of Skibo no longer being a continuing part of the family's life. Perhaps because she had been the only member of the family to have been born at Skibo, she felt a stronger attachment to the old castle than anyone else with the exception of Margaret Miller herself. Migs suggested that her mother should consider giving up her residency in Connecticut and make Skibo her permanent, year-round home. But dear as Skibo was to Margaret, she could not seriously consider permanently leaving her native land. She was an American and would remain so to the end of her life.

If Skibo was to be given to a trust or institution, there were complicated legal and tax issues that must be dealt with and she must seek professional assistance. Several years earlier, Margaret had wisely brought together at Skibo her New York lawyer and friend, John Gray, and her Edinburgh solicitor, Archibald Campbell, so that the two men could discuss the full details of her Scottish holdings, which, in addition to the Skibo estate, included property near Stonehaven.[32] Campbell, since the time of that meeting had retired, but his younger partner and successor in handling Mrs. Miller's affairs, Stephen Seaman, was now a frequent visitor to Skibo in these last summers and was also in close communication with Gray in New York. Fred Mann, executive secretary of the Carnegie Dunfermline Trust, also came up to Skibo, and he and Margaret discussed which items at Skibo, such as the great globe in the library and Andrew Carnegie's desk, might appropriately go to the Birthplace Museum in Dunfermline when she no longer had use for them. Thus the proper steps were being taken in preparation for what must be the inevitable decision of giving up Skibo. The question was no longer if, but only when and precisely how.

In the summer of 1979, Barbara's eldest daughter, Linda, and her husband, Harold Hills, came to Skibo for what they were certain would be their last visit. Linda was eager to experience Skibo to its fullest one more time. This included several picnic lunches in the Fairly Glen. Her grandmother was, of course, no longer able to accompany Linda to the special spot they both loved so very much. But Margaret asked Linda to bring back a bit of the bog myrtle that grew there. This plant, Linda would later write, "has a very distinct, pungent odor, which Grandma said reminded her of the moors. I'll never forget how she sat in the big hall and just closed her eyes and smelled and smelled it, and then asked that it be put in a vase at her bedside. Her greatest pleasures were in the simplest things."[33]

Margaret Miller had once written a short verse to describe her feelings at having to leave Skibo each year:

O dear blue mountains of my home firth,
My heart is sad because I must leave you,
And it will be long before I see you again.

Many years later upon again leaving Skibo for Connecticut, she had copied those remembered lines on the back page of the Skibo Guest Book with the added note: "Written about 1933 & true every year I leave Scotland & Skibo where my heart is. Margaret C. Miller, 1970."[34] Each succeeding year after 1970, upon departing she would add that year's date to the verse—with the terse note after the year 1972, "unavoidably absent." It was the only year she was to miss in the thirty-five summers of her ownership.

In 1980, she once again had Reathel Odum accompany her to Scotland as her companion for the fourth year. They both must have known that this would be the last visit, but the matter was not discussed. They slipped back into the familiar old routine for one more time. There were very few visitors' names entered into the Guest Book that summer. Stephen Seaman came up for a brief business conference in early June, and Migs came to Skibo twice, once in June and again with her daughter, Pamela, for the hunting and fishing in late August. Otherwise, no names were recorded, and so appropriately enough the last name to appear in the eighty-two year roster of distinguished visitors to Skibo was that of Migs—"Aug. 28-Sept. 9th. Margaret M. Flockhart, Englewood, Colo."[35]

The departure from Skibo that year was set for the Thursday after the Dornoch Pipe Band had had its final performance of the year on Saturday, the 13th of September. As always, this final concert of the season would end with a lone piper playing from the top of the turret of the old Dornoch Castle, across the street from the Cathedral. Standing in his Highland dress high above the street and lighted by spotlights from below, the piper would play the traditional Scottish lament of farewell. No summer at Skibo was officially over for Margaret until that final note was sounded. And so it would be in this summer of 1980.[36]

Then followed the usual packing that would occupy the next several days. Margaret tried to avoid giving any special signficance to the preparations for departure this year. Another season was ending and she gave no outward sign to her staff, her family or friends that this year was to be any different. What her own thoughts were, she did not record, and significantly, she did not add the date 1980 to the last page of the guest book. She must have been quite aware, however, that she was now older than either her father or her mother had been when they had said farewell to Skibo for the last time. Unlike either of them, she was voluntarily giving up Skibo. Unsettled as the world might be in the autumn of 1980, she was at least not leaving Skibo under the duress of a world at war and clutching a gas mask under her arm. Of that, at least, she could be thankful.

The farewells to the staff were as usual, quiet and unemotional and as simply stated as had been her acceptance of their welcome to her as their new mistress in 1946. Then as the Mercedes, driven by Pocock, headed down the drive to the north gate, Reathel Odum peered through the back window to catch a final glimpse of dear old Skibo. But Margaret Carnegie Miller had never been one to indulge in lingering farewells. She stared resolutely straight ahead.

CODA

Skibo: Farewell and Hail

1981–1982

Once Margaret Carnegie Miller had made her decision not to return to Skibo after the summer of 1980 and to give the castle and its policies, consisting of some six hundred acres of land immediately surrounding the residence, to the Carnegie United Kingdom Trust, she might well have thought that the issue of what to do with Skibo had been quickly and easily settled, and in accordance with what her parents would have wished her to do. But Andrew Carnegie had once written out of wisdom acquired from his own difficult experiences that "the ways of the philanthropist is hard." Margaret Miller was soon to discover how true her father's words were.

First, somewhat to her surprise, she learned that although the trustees of the United Kingdom Trust were immensely honored to have been chosen as the recipient of Skibo, they regretfully could not accept it as a gift. Margaret had envisioned Skibo's being converted by the Trust into a great educational center to promote free inquiry and the progressive enlightenment of British society—something akin to the Princeton Institute for Advanced Study or Robert Hutchins's Center for the Study of Democratic Institutions in California—a place where leading scholars of Britain and of the world could come to pursue their own work and exchange ideas in the peaceful and isolated environment that was Skibo's. It was a noble idea, one that certainly had appeal to the trustees and was consonant with the Trust's mission. But Geoffrey Lord, the Trust's executive secretary, had to inform Mrs. Miller's lawyers that unfortunately the annual income from endowment would not provide the funds sufficient to maintain the Castle and its grounds, even if the Trust's charter should allow its funds to be used in this manner. To give but one example, the minimum heating required to keep the pipes from freezing in Skibo, which had no one in residence during the winter months, cost something over £17,000 a year. To carry on a quality program of symposia and conferences and to support scholars in residence would absorb a major part of the Trust's total annual budget, leaving little else for its other activities.

The United Kingdom Trust, however, did employ the Edinburgh architectural firm of Law and Dunbar Naismith to make a study of potential future uses for Skibo Castle and estimate the cost to convert the building for each of these possible functions. The architects did a thorough study and in their detailed report submitted to the United Kingdom Trust several months later, there were some interesting, even exotic, suggestions.

The easiest, least expensive, and least innovative possibility would be to keep Skibo as a private residence. This would not conflict with any zoning restrictions nor affect its national registry status on the official listing of the Scottish Development office. It could be offered to some wealthy individual for a country home. Any expenses incurred for repair and renovation would be the new owner's responsibility, and the United Kingdom Trust could add the purchase price obtained to its existing endowment. The report offered the hopeful information that only recently "an Arab had purchased property in Eastern Ross and Skibo is a better bargain."

Another suggestion was to convert Skibo into a luxury class hotel. Dornoch was renowned throughout Britain for its mild climate, and its fine Royal Golf Course and sand beaches. There were also the excellent hunting and fishing opportunities on the estate itself. The Castle could be quite easily converted into a hotel for approximately £325,000. If in addition, as the report highly recommended, small cloistered cottages or chalets should be built in the wooded areas on the north shoreline of Loch Ospisdale to be rented for the summer, the cost would be considerably higher, but the returns would be even greater. The report even made the improbable suggestion that "the Club Mediterranean should not be overlooked" as a possible client. Any zoning difficulties related to Skibo's commercial use could be satisfactorily resolved, it was believed, because the "Regional Council and Highlands and Islands Development Board would strongly encourage tourist related development to inject capital into a depressed local economy and provide long term employment for the local population."

Another whole range of possibilities lay within the area of institutional use. Skibo would make a beautiful residential preparatory school or college, but the cost of conversion would be high, at least two million pounds, and the number of clients interested in obtaining Skibo for this use would be very limited. Either the national or local government might be interested in Skibo as a public institution for a mental institution, a convalescent home or a minimum security prison. But its remote location, making it difficult for families and friends to visit the patients or inmates, militated against these latter institutional possibilities. There was even the fascinating suggestion that NATO might be interested in Skibo, precisely because of its isolated location, as a place to establish its Defense Headquarters, although the authors of the report had to admit that this was "a remote possibility." If it should prove impossible to find a prospective client for any of these functions, "then the sensible course would be to apply [to the Scottish Development Board] for demolition instead of allowing the castle to become dilapidated." But this latter option should only be considered as a last resort, and the report concluded on the upbeat note that it should be "feasible to convert the castle. . . . We are by no means pessimistic about securing a suitable use. We were encouraged that all authorities contacted welcomed [our] initial report."[1]

If Margaret Miller ever read this report, she must have been both amused and dismayed by some of the suggestions offered. Imagine her beautiful Skibo's being used as an insane asylum or a prison! And what would Andrew Carnegie, the ardent pacifist, have thought of his Skibo as the defense headquarters for NATO? No—there was only one viable possibility if the United Kingdom Trust could not itself use Skibo as part of its educational program. Skibo must be sold to a private individual and the funds thus obtained could then be used to further the work of the Trust.

But even if one assumed the U. K. Trust could find a purchaser, there were still some very complicated legal difficulties involved in Margaret Miller's transferring Skibo to the trust and in meeting the tax obligations that might be involved. Andrew Carnegie has often been praised, and quite rightly so, for having pursued his own Gospel of Wealth in an age when there was little or no tax incentive to do so. He gave away nine-tenths of his great fortune during his life and prior to the time there was a federal income or inheritance tax to spur on his benevolent intentions. After World War I, with both federal and state income and inheritance taxes steadily rising to ever higher levels, many other men of great wealth, and many corporations as well, became ardent new disciples of Carnegie's old Gospel. Philanthropy—the love of mankind—hardly seems the appropriate word to apply to the many new, hastily created "charitable" foundations. We need another word, perhaps misotely—the hatred of taxes—would do, to describe their reason for being. It is hardly surprising that in time the people and their governments should begin to question the motivation for creating many of these new foundations and to respond with legislation designed to check some of the more flagrant abuses. The Internal Revenue Service in the United States began to supervise closely the annual audit sheets of these tax sheltered foundations to make sure their return on endowment was actually being spent and spent in line with the foundations' stated benevolent purposes. Great Britain took the even more draconian step of placing a limit of £250,000 that any individual or corporation could give tax free for philanthropic purposes. Thus had government through its taxing powers both provided an incentive for giving that had not existed in Carnegie's day and at the same time, ironically enough, had made it much more difficult to do so.

Carnegie's daughter in 1982, in trying to pursue her father's philanthropic ideals, found herself in a particularly difficult position. She was an American citizen, and consequently subject to American tax laws. But the property she sought to give to a long established and highly reputable philanthropic trust was in Scotland, and the Trust which was to be the recipient was itself chartered in Great Britain. She, because of her four month's residence in the United Kingdom each year, was also considered to have legal domicile in that country and was subject to its tax laws and its restrictions on giving. Both her New York and her Edinburgh lawyers, as well as her grandson-in-law, Louise Thomson Suggett's husband, Gavin, whom Margaret Miller had asked to participate in the negotiations as a representative of the family, were to spend many long months wrestling with the thorny problems this intended act of generosity had raised. There must have been moments when Margaret Miller felt that the simplest answer would be that gloomy "final solution" which the architects' report had suggested might prove necessary—that is, simply level Skibo to the ground and turn its policies into a park and wild life sanctuary. Sometimes this seemed preferable in any event to the alternative of having some stranger owning and living in her Skibo. But these were only passing thoughts, born out of her sorrow in having to give up Skibo and her frustration in trying to do so in the most useful and benevolent way. There was no recourse but to put one's faith in the lawyers' finding a solution that would somehow satisfy the legal constraints and tax obligations imposed by the two countries involved. It was finally determined in the late spring of 1982 that the only possible procedure would be for Margaret Miller herself to sell Skibo Castle and its policies, pay the taxes due both countries out of the proceeds of the sale and then give the remainder that was left after paying all the necessary taxes to the Carnegie United Kingdom Trust.[2]

In the meantime, Savills of York, London and Edinburgh, and a local realtor, Renton Finlayson of Bonar Bridge, Sutherland, were given the commission as joint selling agents to list Skibo as being on the market for sale. Margaret Miller had also decided to sell, in addition to the Castle itself and its policies for the benefits of the United Kingdom Trust, almost all of the remainder of the Skibo estate, the proceeds of which would be added to her own personal holdings. Only certain timber lands of the Skibo estate would be withheld from the sale for later disposition. She also made the qualification, out of respect for the wishes of her family and particularly for her daughter, Migs, that even though she was selling the fishing and shooting rights of the Evelix area, along with the lands, loch, and stream, she reserved for the next twenty-five years the right for herself, "her successors and her guests whether accompanied by her or not, and any others authorized by her, or her Agents in the United Kingdom, to fish at any time for any fish in season, the lease being for a maximum of two rods and to have first choice of where those rods shall fish on any particular occasion. The maximum number of rods allowed to exercise the fishing rights at any one time shall be restricted to four on the loch [Evelix] with a further two on the river for the period of the lease. The Seller will have first call on the shootings for any period of not more than 21 consecutive days during the season, 1st September to 1st February in any year during the period of the lease."[3]

The tenants of the Skibo farms and crofts and the villagers of Clashmore, Spinningdale, and Bonar Bridge had their first real confirmation of the fact that the long Carnegie-Miller era was over when the word was quickly circulated throughout the parishes in the spring of 1981 that Mrs. Carnegie Miller would not be returning to Skibo that summer. Soon the Scottish press was spreading the story throughout Britain that the golden thread that had for so long tied Skibo to the Carnegie steel fortune in America had been snapped. Feature writers poured out their stories, highly flamboyant and largely inaccurate, of the Carnegie glory days to titillate their readers with romantic fantasies of that Edwardian age of opulence which was by now, after two World Wars and a great depression, as romantically remote as was the Elizabethan age of swashbuckling adventure. Other journalists, exploiting a more populist sentiment, raced to Dornoch and the little village of Clashmore to interview the local citizenry and report on their distress over the future. "Shadow Over Clashmore" screamed the headline above the feature story by George Birrell in the *Daily Express* of August 1981. Birrell pictured Clashmore as being now a ghost town of twenty-five inhabitants, waiting in fearful uncertainty to learn what might happen to them and to their village. "Her [Margaret Miller's] absence has brought the chill wind of economic reality to an estate which was always protected from the vagaries of such things. . . . Carnegie would always boast he could fish from his own lochs, drink milk from his own cows, eat grapes from his greenhouses, swim daily in his private pool, golf on his own course and shoot over his own moors. That is the lifestyle we now associate only with the wealthy Dutchmen or Arabs who have bought properties in Scotland. But they rarely display the benevolence of Carnegie . . . and that's what the locals fear most."[4]

In September 1981 the story was released that Margaret Carnegie Miller had given Skibo Castle to the Carnegie United Kingdon Trust, but the concern of the residents of the region for the future was hardly alleviated by this news, for the same issue of the newspapers also carried the story that the trustees would not be able to keep and maintain the estate and would have to put it on the market for sale. Geoffrey Lord, the secretary of the Trust, was quoted as saying that the trustees did understand and sympathize with the people of the region. "The trustees recognise the anxieties in the local community about the future of the Castle and the estate. When large properties built in more prosperous times have to be sold it is usually a matter of anxiety especially for those who have links with the estate." But Lord, on behalf of the trustees, expressed the hope that the sale would be of benefit to the community and the realization of these assets, resulting from the sale "should provide a substantial addition to the trust's present endowment, and thus allow the trustees to carry on more effectively the wishes of their founder."[5]

In May 1982, the tax and legal issues having been resolved to an extent sufficient to allow Skibo to be placed on the market, the Savills advertisement was placed in the newspapers. A handsome prospectus of the estate was also issued listing sixty separate lots that would be sold either as individual units or in blocks. Except for the Castle, its policies and sea frontage, each lot carried a suggested price, which was to be considered as only a guide to prospective bidders and not as an established and firm cost for that piece of property. The lots ranged from the Castle and its policies, consisting of eighty-five acres, and the home farm of 685 acres, with a guide price for the latter of £500,000, down to the Fairy Glen lot consisting of six acres, currently being rented for £5 per year, with a guide price of £300. It was a painful wrench to find Fairy Glen, with its priceless memories, being offered at the lowest price of all the lots. In all, 19,000 acres of the 22,000-acre estate was being offered for sale, with a total suggested guide price of £1,855,600, exclusive of the castle, its policies, and the lochs and sea frontage.

The prospectus contained the statement that "for the avoidance of doubt it has been decided that unless sold beforehand, there will be a closing date for best offers in Scottish Legal terms to be submitted to Savills office in Edinburgh by noon Tuesday 27th July [1982]."[6] Soon after the prospectus appeared, however, Margaret Miller withdrew three of the farms, Fload, consisting of 263 acres, Acharry, 100 acres, and Creich, 362 acres, from the sale. Three of Lord Migdale's daughters, Margaret and Mary Thomson and Louise Suggett, had formed a partnership and by pooling their resources were able to purchase these farms as a unit from their grandmother. It pleased Margaret Miller immensely that a part of Skibo, consisting of three of its best farms, was to remain within the family.[7] Margaret and Mary were close by at Ospisdale and with Margaret's keen interest in farming, this should prove to be a most happy and successful arrangement.

The people of Dornoch and Creich parishes were also cheered by this news of a continuing family interest in a part of the Skibo estate, but all through the long spring and much of the summer they must now wait to learn of the disposition of the Castle and the remainder of the estate, and of their own future which was so inextricably tied to that of Skibo. Wild speculations and rumors of Arab sheiks landing by helicopter on the castle grounds to inspect the property were transmitted throughout the region with a celerity that approached the speed of light. Most of the inhabitants, with the typical Scottish pessimism of expecting the worst, were convinced that only an Arab oil tycoon, or a Japanese or Dutch syndicate, could possibly afford to purchase the estate. At a gathering of the people of the area on the grounds of Skibo in September 1981, Margaret Miller's good friend and minister, James Simpson, was asked to offer a few remarks which might be of consolation in this moment of painful transition. After giving a brief statement on the years of the Carnegie-Miller tenure, he concluded with:

> Margaret's dream for her beloved Skibo is that somehow in God's scheme of things this castle, or the money which the castle will make available, will be used to enrich the lives of others. I am sure it is the earnest prayer of all of us gathered here that her wish will be fulfilled.
>
> Let us pray. Father we thank you for this home, set in such a lovely part of your world. We thank you for all it has meant to the Carnegie family these past eighty years. We thank you for the happy times that Mr. and Mrs. Carnegie spent here, for the equally happy hours that Margaret knew. We thank you for all who came to this castle and left refreshed and recreated.
>
> Father, one chapter in the history of this castle is ending. Another chapter is about to begin. Father, to those who will soon be responsible for this castle, grant your divine wisdom, that through the mystery of your providence this castle may continue to enrich people's lives.
>
> And Father, remind each one of us gathered here today that the great use of life is to use it for a purpose that will outlast it, that he who would be great in your sight must become the servant of all.
>
> For Christ's sake, we ask it.[8]

To this prayer the people of Dornoch could offer a most ardently reverential Amen.

At last, the long-awaited day of 27 July arrived. In the Savills office in Edinburgh, the bids for Skibo Castle, its policies, and the other lots of the estate were opened. Quickly the news was conveyed to Sutherland County. The new laird of Skibo Castle, its policies and most of that which remained of the 19,000-acre estate would be Derek Holt, head of Holt Leisure Parks, Ltd., with headquarters in Renfrewshire. His best offer was not disclosed, but it was thought by the press to be approximately £2.5 million.[9] The new laird would be one of their own, after all, and every true Scotsman could rejoice in that.

Derek Holt had been born in near-by Inverness, so the stories of the legendary Skibo had been a part of his childhood heritage. As a young boy, however, he had moved south with his parents to a farm in Lancashire, which he eventually inherited. He proved to be a successful farmer, and by the early 1960s he began to seek potential growth areas in which to invest his surplus capital. Believing that there were great possibilities for profit in the growing demand for recreational facilities, he had opened a caravan (trailer) park in northern England. The success of this venture enabled him to establish other such camping grounds in Scotland, and soon thereafter he returned to his native land, where he bought a home in Renfrewshire for himself and his family. His fortunes truly escalated a short time later with his opening of a marina in Inverkip—the first real marina in all of Scotland. Within ten years, he was in a financial position strong enough to outbid all other aspirants, both domestic and foreign, for the possession of Skibo.

Why, he was asked by the reporters, had he sought to buy Skibo. "Well," he replied, "it was a desire to get back to the land and farming. My daughters, who are in the family business with me, wanted to return closer to the land. Skibo Castle seemed the perfect answer and it also gives us a challenge in our new line of business." He quickly added that this "new line of business" did not mean that Skibo would be turned into a commercial hotel. "It will be our home but to pay the bills we hope to develop it more as a luxury guest house where people can enjoy traditional Highland sports of prime fishing and hunting." He also assured the public that he had no intention of drastically altering the castle or diminishing its historical association. "It will always be Andrew Carnegie's house," he promised his interviewers.[10]

Here was a happy ending to the long months of suspenseful waiting. The Regional Council and the Highlands and Islands Development Board were as pleased with the results as if they themselves had been given the opportunity to select the new laird from the long list of applicants. Holt's plans for Skibo would mean a boost to tourist trade in northern Scotland and precisely that "injection of capital into a depressed local economy," which these administrations had been quoted as seeking in the report made by the architects on Skibo's potential uses. And as for the parishioners of the region, they surely must now be convinced that the prayers of their good minister, James Simpson, had indeed been answered.

In expressing pleasure over obtaining this particular new laird for Skibo, the people of the region were simply recognizing that the best hope of economic prosperity for northern Scotland now lay with the development of tourism and the promotion of recreational facilities. In so doing, they were but reflecting the realities of the age in which they lived. Indeed, one might do well to look to the biographies of those who have possessed Skibo over the last one thousand years as a source for understanding the changes that have occurred in the concentration of wealth and power in the Western world in this past millennium. It is a long road through history that one must journey—from the Norse Vikings who built their fortress on the lands of Skibo in the 980s to protect their long prowed boats of conquest, down to those who build marinas for small, plastic bottomed sailing craft and sleek, teak yachts of pleasure in the 1980s. But always Skibo has gone to the economic victors of any particular age. In the disorder and chaos that had followed the disintegration of the Roman empire, power belonged to the physically muscular, wealth was a product of rapine and the Vikings had come to Skibo. Later, when a new source of power and order had emanated from Rome, the Church became the single greatest monopolizer of wealth, and the good bishop of Caithness, Gilbert de Moravia, and his successors in the diocesan chair, were the proud and rightful possessors of Skibo. But Roman Catholicism in Britain was to prove to be as transitory as Roman Caesarism, and in the storm of the Protestant Reformation, its wealth, like its ecclesiastical authority, was to pass to the Protestant landed gentry and to their legislative bodies. For the next four hundred years it would be the Grays, the Dempsters, and the Sutherland-Walkers who would possess Skibo. The late nineteenth and early twentieth century, as George Dempster had anticipated, would belong to the industrialists—the manufacturers of textiles and steel, and Skibo would belong to Andrew Carnegie and be exalted to a higher glory than it had ever known before. But now in the last decades of the twentieth century, in this age of triumphant consumerism when there was more profit in building Winnebago caravans than in manufacturing rails and locomotives, in owning marinas and ski resorts than in possessing textile mills in Manchester or steel mills in Pittsburgh—ah then, Skibo would open her gates to the president of Holt Leisure Parks, Ltd. Skibo's history provides a fascinating lesson in microeconomics.

In far distant Connecticut, the last representative of Skibo's old order had also been waiting eagerly to hear who would succeed her at Skibo. And when the news came, she too was delighted. Margaret Carnegie Miller was particularly pleased to be informed that June Holt, the new lady of Skibo, was an enthusiastic horticulturist. One of Skibo's most attractive features to the Holt family had been its beautiful gardens. Margaret could be assured that they would continue to be well tended, and for that she was grateful.[11]

Margaret Miller had need for some good news at this time, for on the very day that the sale of Skibo was consummated, she received some very sad news. Her youngest daughter, Migs, had had rather serious open heart surgery in the late spring of 1982. The operation was successful, but in early July she became ill with hepatitis, which she had apparently contracted from the blood transfusion given during surgery. She bravely fought the disease for several weeks, but on 24 July, the very day the bids on Skibo were opened, she died. Migs had been the one member of the family who had never become totally reconciled to the idea of giving up Skibo. For a time, she had even briefly entertained the idea of putting in a bid for one of the cottages on the estate and if successful, making it her summer home. But the reality of double taxation by becoming a resident of both Britain and of the United States, had, of necessity, dissuaded her. The tragic coincidence of her death's occurring on that very day that Skibo was sold could not fail but be noted by her family and friends. It was almost as if Migs's heart could not withstand the loss.[12]

In early September, the season at Skibo which Migs had always most preferred, one of her daughters, Gail, and husband, took her ashes back to Skibo and scattered them on the Loch Evelix. Migs's other daughter, Barbara, was in Hawaii and unable to accompany her sister, but she had sent by air express a lei of white orchids. This too was cast into the loch. Exotic as these blossoms were to Scotland, they nevertheless floated easily like miniature white waterlilies on the surface of the water as the current carried the garland out to sea—a last farewell to the lands and waters of Skibo.[13]

So the second millennium of Skibo would begin with a new laird. The people of the old estate stood ready to hail the advent of still another era in the land's history. Margaret Carnegie Miller was confident that the Holts would quickly discover the truth of that motto which Andrew Carnegie had once had hung in the great hall, "Hame is hame, but a hieland hame is mair than hame." And even that was only half of the truth, for Skibo, to its long line of lairds and their families, had always been even mair than a hieland hame. The first Celtic inhabitants had known there was something special here and they had named the place "Schyth"—the fairyland of peace. The slow change of that name into Skibo had in no way dispersed its old magic.

So "Slàinte" to old Skibo and to all who had lived and loved there. And for those who will come later, may you know tight lines, hot barrels and above all, peace—that peace which some believe was the gift of the Gaelic fairies, that peace which Nature has so clearly intended for this beautiful land and all of its good people.

Notes

CHAPTER I

1. See David Dorward, *Scotland's Place Names* (Edinburgh: William Blackwood, 1979), p. 42.
2. *Ibid.*, p. 41.
3. Quoted in *Dornoch*, a most useful little pamphlet, although no author's name is given. (Dornoch: Dornoch Craft Centre, Ltd., no date of publication), chap. 4, no pagination.
4. For the text of the letters of legitimation and a full genealogy of the Gray family of Skibo see Peter Gray, *Skibo: Its Lairds and History* (Edinburgh: Oliphant, Anderson & Ferrier, 1906), p. 19 ff.
5. *Ibid.*, p. 42.
6. Oliver Goldsmith, "The Deserted Village," reprinted in the *Harvard Classics* series, ed. by Charles W. Eliot (New York: P. F. Collier & Son, 1910), Vol. 41, pp. 522–23.

CHAPTER II

1. The deed is quoted in W. MacGill's article, "Side-Lights on Old Times," *Glasgow Herald*, 11 February 1905, and reprinted in Peter Gray, *op. cit.*, pp. 37–38.
2. For a more detailed account of the Dempsters' spinning and weaving ventures, see William Calder, *The Estate and Castle of Skibo* (Edinburgh: The Albyn Press, 1949), p. 32.
3. For an account of Dempster's agricultural and fishing improvements, see the biographical sketch on George Dempster by Robert

Harrison in *The Dictionary of National Biography*, Vol. V (Oxford University Press, 1959–60), pp. 784–85.

4. Calder, *op. cit.*, pp. 31–32.
5. *Ibid.*, p. 33.
6. Edwin Arlington Robinson, "Miniver Cheevy," as published in Louis Untermeyer, ed., *Modern American Poetry* (New York: Harcourt, Brace and Company, 1942), p. 140.

CHAPTER III

1. Andrew Carnegie, *Autobiography of Andrew Carnegie* (Boston: Houghton Mifflin Company, 1920), p. 217.
2. Andrew Carnegie's response to a toast given by Lord Rosebery at a dinner held in Edinburgh, 8 July 1887, quoted in Burton J. Hendrick and Daniel Henderson, *Louise Whitfield Carnegie* (New York: Hastings House, 1950), pp. 98–99.
3. Louise Whitfield Carnegie to her mother, Fannie Davis Whitfield, 17 July 1887, quoted in Hendrick and Henderson, *op. cit.*, pp. 101–2.
4. Louise Whitfield Carnegie to her mother, Fannie Davis Whitfield, 10 June 1887, quoted in Hendrick and Henderson, *op. cit.*, p. 95.
5. Louise Carnegie to Fannie Whitfield, – July 1888, quoted in Hendrick and Henderson, *op. cit.*, p. 123.
6. Louise Carnegie to Fannie Whitfield, no date given as quoted in Hendrick and Henderson, *op. cit.*, p. 129.
7. Andrew Carnegie to Louise Carnegie, no date given as quoted in Hendrick and Henderson, *op. cit.*, p. 145.
8. Louise Carnegie to Andrew Carnegie, no date given as quoted in Hendrick and Henderson, *op. cit.*, p. 147.
9. Louise Carnegie to Charles H. Eaton, no date given as quoted in Hendrick and Henderson, *op. cit.*, p. 147.
10. Andrew Carnegie to Hew Morrison, 18 December 1900, Andrew Carnegie Papers, Library of Congress (hereafter ACLC), Vol. 80.
11. Andrew Carnegie to Hew Morrison, 11 March 1902, ACLC, Vol. 88.
12. Typed manuscript of Skibo Library Catalogue by Hew Morrison in the Skibo Castle Library.
13. This is an often repeated story. See particularly, Calder, *op. cit.*, p. 72.
14. Louise Carnegie to Charles H. Eaton, no date given as quoted in Hendrick and Henderson, *op. cit.*, p. 152.

15. Speech of Andrew Carnegie to the tenants of Skibo, no date given as quoted in Hendrick and Henderson, *op. cit.*, p. 151.
16. Andrew Carnegie to George Lauder, 23 June 1899, ACLC, Vol. 66.
17. Andrew Carnegie as quoted in Burton J. Hendrick, *The Life of Andrew Carnegie* (Garden City, N.Y.: Doubleday, Doran, 1932), Vol. II, p. 88.

CHAPTER IV

1. Andrew Carnegie to George Lauder, 26 February 1901, ACLC, Vol. 82.
2. J. P. Morgan as quoted in Frederick Lewis Allen, *The Great Pierpont Morgan* (New York: Harper Bros., 1949), p. 134.
3. J. P. Morgan as quoted in Hendrick, *Life of Andrew Carnegie*, Vol. II, pp. 138–39.
4. Andrew Carnegie to John Morley, 3 February 1901, ACLC, Vol. 81. For a full account of the sale see Joseph Frazier Wall, *Andrew Carnegie* (New York: Oxford University Press, 1970), Chapter XX, "The Contract Closed, 1901."
5. Much of the remainder of this chapter, some of it verbatim, has been taken from Chapter XXIV, "Sunshine at Skibo, 1901–1914," of my biography of Andrew Carnegie, *op. cit.* I have considered it unnecessarily pretentious to enclose within quotation marks those paragraphs in which I quote myself directly and which I include here with the kind permission of Oxford University Press.
6. Andrew Carnegie to George Lauder, no date given as quoted in Hendrick, *Life of Andrew Carnegie*, Vol. II, p. 154.
7. See Angus Macpherson's delightful autobiography, *A Highlander Looks Back* (Oban: Oban Times Limited, 1955), pp. 31–33.
8. Andrew Carnegie to John Morley, 5 August 1908, ACLC, Vol. 155.
9. John Morley to Andrew Carnegie, 8 August 1908, ACLC, Vol. 155.
10. See Hendrick and Henderson, *op. cit.*, pp. 176–77, for an account of King Edward VII's visit.
11. Andrew Carnegie to John Morley, 5 August 1908, ACLC, Vol. 155.
12. George Lauder, Sr., to Louise Carnegie, 18 August 1895, quoted in Hendrick and Henderson, *op. cit.*, pp. 127–28.
13. Interview with Margaret Carnegie Miller, 29 March 1982.

14. Andrew Carnegie to James Donaldson, 20 March 1904, ACLC, Vol. 104.
15. Andrew Carnegie to John Morley, 29 June 1904, ACLC, Vol. 105.
16. Andrew Carnegie to John Morley, undated, ACLC, Vol. 164.
17. Andrew Carnegie to R. W. Gilder, 2 August 1906, ACLC, Vol. 132.
18. Andrew Carnegie to John Morley, 24 November 1905, ACLC, Vol. 122.
19. Andrew Carnegie to John Morley, 14 January 1906, ACLC, Vol. 124.
20. Andrew Carnegie to John Ross, 11 February 1913, ACLC, Vol. 213.
21. Andrew Carnegie, in a speech at the Chambers Institute in Edinburgh, as printed in his pamphlet, *William Chambers* (Edinburgh, 1909).
22. Andrew Carnegie, *Autobiography* (Boston: Houghton, Mifflin Company, 1920), pp. 371–72.
23. Andrew Carnegie to John Ross, 17 August 1914, ACLC, Vol. 223.
24. Andrew Carnegie, "A League of Peace–Not 'Preparation for War,'" *Independent*, Vol. 80, 19 October 1914, reprint in ACLC, Vol. 226.
25. Louise Carnegie's diary entry, 22 April 1919, quoted in Hendrick and Henderson, *op. cit.*, p. 212.
26. As quoted by Robert Morrison, Carnegie's valet, in an interview, August 1959.
27. John Morley to Louise Carnegie, undated, ACLC, Vol. 239.
28. This cairn cannot be seen from the road, but any of the local inhabitants is pleased to point it out to visitors.

CHAPTER V

1. Louise Carnegie to Stella Whitfield, quoted but without giving a date, in Hendrick and Henderson, *op. cit.*, p. 224.
2. Louise Carnegie to Louise Miller Thomson, 17 June 1939, reprinted in Hendrick and Henderson, *op. cit.*, p. 223.
3. Louise Carnegie to Margaret Miller, 30 July 1920, quoted in Hendrick and Henderson, *op. cit.*, p. 229.
4. *Ibid.*
5. Louise Carnegie to Margaret Miller, as quoted, with no date given in Hendrick and Henderson, *op. cit.*, pp. 230–31.
6. *Ibid.*, also undated, pp. 231–32.

7. Louise Carnegie to Margaret Miller, 25 September 1920, as reprinted in Hendrick and Henderson, *op. cit.*, pp. 233–34.
8. *Ibid.*, p. 233.
9. Skibo Castle Guest Book, for the year 1921, in possession of the Carnegie Dunfermline Trust, Dunfermline, Scotland.
10. Louise Carnegie to Margaret Miller, quoted, but no date given, Hendrick and Henderson, *op. cit.*, p. 235.
11. Louise Carnegie to Margaret Miller, quoted, but no date given, in Hendrick and Henderson, *op. cit.*, pp. 235–36.
12. Louise Carnegie to the Millers, quoted, but no date given, in Hendrick and Henderson, *op. cit.*, pp. 237–38.
13. *Ibid.*, p. 241.
14. Diary entry by Louise Carnegie, 28 July 1938, quoted in Hendrick and Henderson, *op. cit.*, p. 287.
15. Hendrick and Henderson, *op. cit.*, p. 288.
16. Margaret Carnegie Miller, *Diary, 1939–1946*, Vol. I, p. 69. This remarkably fine diary is in the possession of Mrs. Miller and is used with her kind permission.
17. *Ibid.*, p. 7.
18. *Ibid.*, pp. 20–21.
19. *Ibid.*, p. 16.
20. *Ibid.*, p. 29.
21. *Ibid.*, p. 33.
22. *Ibid.*, p. 52.
23. *Ibid.*, pp. 53–56.
24. *Ibid.*, pp. 49–50.
25. Margaret C. Miller, *Diary*, Vol. II, pp. 28–29.
26. *Ibid.*, p. 13.
27. Louise Carnegie to Louise Thompson, quoted, but no date given, in Hendrick and Henderson, *op. cit.*, pp. 294–95.

CHAPTER VI

1. The story of Margaret Carnegie Miller's return to Skibo in 1946 and the direct quotations of what was said at that time are taken from a manuscript of her memories of Skibo which Mrs. Miller prepared for me in 1982. It is hereafter referred to as M. C. Miller, "Memories of Skibo."
2. The following material on hunting and fishing in Scotland was provided by Harry Blythe in an interview that I had with him at his retirement home, Haugh Court, in Inverness, Scotland, 2 June 1982.

3. Interview with Blythe, *loc. cit.*
4. Entry of 11 August 1947 in the *Hunting Log Books of Roswell Miller III*, Vol. I, in the possession of his son, Kenneth Miller, who kindly gave permission to me to use.
5. *Ibid.*
6. The following material on the family was provided me by Gordon Thomson (Lord Migdale) in a conversation at his home, Ospisdale, 31 May 1982. Lord Migdale was most helpful in his discussion of the Miller family relationships in the years after 1947, and I am most appreciative of the insights he has given me.
7. Conversation with Louise Thomson Suggett at her home in Kilspindie, Perthshire, Scotland, 3 June 1982.
8. Conversation with Elizabeth Thomson Milligan at her home in Edinburgh, 7 June 1982, and with William Gordon Thomson in London, 8 June 1982.
9. Conversation with Margaret Thomson at her home, Ospisdale, 31 May 1982.
10. Linda Thorell Hills to Joseph F. Wall, 20 July 1982.
11. Excerpt from Linda Thorell's diary, dated 1 October 1967, kindly sent to me by its author.
12. Linda Thorell Hills to Joseph Wall, 20 July 1982.
13. *Ibid.*
14. Entry of 14 September 1947, *Hunting Log of Roswell Miller III*, Vol. I, *loc. cit.*
15. Miller, *Hunting Log*, Vol. III, *loc. cit.*
16. Interview with J. R. Whittet at Skibo Castle, 2 May 1958.
17. Miller, *Hunting Log*, Vol. III, *loc. cit.*
18. Interview with Mr. and Mrs. Bobby Edgar at their home near Spinningdale, Scotland, 1 June 1982. Neal Campbell, the factor appointed in 1981 by Savills, the selling agents for the Skibo estate, is also critical of McLeish's tenure as factor. Although not knowing McLeish personally, Campbell was critical of the fact that the permanent records on file in the Estate Office were not properly kept up after 1964 and that few incoming letters to the factor were answered. Interview with Neal Campbell, Skibo Estate Office, Clashmore, Scotland, 31 May 1982.
19. Interviews with Lord Migdale, *loc. cit.*, 31 May 1982, and with Louise Thomson and Gavin Suggett at their home, Kilspindie, Perthshire, Scotland, 3 June 1982.
20. Interview with Stephen Seaman in his law office, Edinburgh, 3

June 1982, and with John Gray in his home, Greenwich, Connecticut, 30 September 1982.

21. M. C. Miller, "Memories of Skibo."
22. Diary of Reathel Odum, Margaret Miller's companion at Skibo for the summers of 1977–80. Entries in that diary for 21 July and 25 August 1977, and also M. C. Miller, "Memories of Skibo."
23. M. C. Miller, "Memories of Skibo."
24. *Ibid.*
25. *Ibid.*
26. Conversations with Lord Migdale; Kenneth Miller, Margaret and Roswell Miller's grandson; and Reathel Odum, Margaret Miller's companion.
27. Ruth Adams to Joseph Wall, 18 April 1982. Miss Adams's letter was most helpful in providing me with information on the daily routine at Skibo. She also sent me a complete menu for every meal served at Skibo during her two-week stay.
28. Alice G. Hoskins to Reathel Odum, 4 January 1977. In this letter, Mrs. Hoskins describes the position and inquires as to Reathel Odum's possible interest in being recommended for it. Letter in the personal papers of Reathel Odum, Seabrook, St. John's Island, South Carolina.
29. I am greatly indebted to Reathel Odum and to John Snyder for their kindness in permitting me to read Miss Odum's letters and her diary. Much of the following material in this chapter has been taken from these sources. These papers are in the possession of Reathel Odum, Seabrook, St. John's Island, South Carolina.
30. See letter of Ruth Adams to Joseph Wall, 18 April 1982, and scattered entries in diary of Reathel Odum for the year 1977.
31. Entry for 17 May 1978, diary of Reathel Odum, *loc. cit.*
32. M. C. Miller, "Memories of Skibo."
33. Linda Thorell Hills to Joseph Wall, 20 July 1982.
34. *Skibo Guest Book, loc. cit.*
35. *Ibid.*
36. Reathel Odum to John W. Snyder, 4 September 1980, letter in the possession of John W. Snyder, Seabrook, St. John's Island, South Carolina.

CODA

1. *Skibo Castle: Report on Potential Uses*, prepared by Law and Dunbar Naismith, Architects, Edinburgh, Scotland. The report is

in the files of the Carnegie United Kingdom Trust, Comely Park House, Dunfermline, Scotland. The use of this report and all of the direct quotations from the report given in the paragraphs above are with the kind permission of Geoffrey Lord, Secretary of the Carnegie United Kingdom Trust.

2. The above information on the legal and tax problems involved in disposing of Skibo for the benefit of the Carnegie United Kingdom Trust was obtained from some very helpful conversations with Stephen Seaman in Edinburgh, 3 June 1982, Gavin Suggett in Kilspindie, 3 June 1982, and John Gray in Greenwich, Connecticut, 30 September 1982.
3. Prospectus for *Skibo Estate,* prepared by Savills and Renton Finlayson, p. 24.
4. George Birrell, "Shadow Over Clashmore," *Scottish Daily Express* (Glasgow), 7 August 1981.
5. James Johnston, "Trust Given Carnegie's Castle Home," *Scottish Daily Express* (Glasgow), 22 September 1981.
6. *Skibo Estate* prospectus, pp. 3–5 and Appendix of Closing Date and Guide Prices.
7. Conversation with Louise and Gavin Suggett, 3 June 1982.
8. Manuscript of the Rev. James Simpson's remarks at Skibo, September 1981, in the possession of Mrs. Margaret Carnegie Miller and used with her kind permission.
9. *The Scotsman* (Edinburgh), 27 July 1982.
10. *The Scotsman* (Edinburgh), 29 July 1982.
11. Conversation with Mrs. Margaret Carnegie Miller, 30 September 1982.
12. Conversation with John Gray, 30 September 1982.
13. Conversation with Margaret Carnegie Miller, 30 September 1982.